Christine Rösch, Klaus-Rainer Bräutigam, Jürgen Kopfmüller,
Volker Stelzer, Patrick Lichtner, Annika Fricke

Indicator-based Sustainability Assessment of the German Energy System and its Transition

Indicator-based Sustainability Assessment of the German Energy System and its Transition

by
Christine Rösch, Klaus-Rainer Bräutigam, Jürgen Kopfmüller,
Volker Stelzer, Patrick Lichtner, Annika Fricke

KIT Scientific Publishing

Covergestaltung basiert auf Freepik
Icons: hands and car made by Freepik from www.flaticon.com

Impressum

KIT Scientific Publishing

Karlsruher Institut für Technologie (KIT)
KIT Scientific Publishing
Straße am Forum 2
D-76131 Karlsruhe

KIT Scientific Publishing is a registered trademark
of Karlsruhe Institute of Technology.
Reprint using the book cover is not allowed.

www.ksp.kit.edu

Print on Demand 2018 – Gedruckt auf FSC-zertifiziertem Papier

ISBN 978-3-7315-0792-5
DOI 10.5445/KSP/1000082161

Foreword

Within the Helmholtz Alliance ENERGY-TRANS, a large third-party funded project coordinated by KIT-ITAS, an indicator-based Sustainability Indicator System (SIS) was developed. The research in the Alliance ENERGY-TRANS focused on the energy transition and related requirements for the transformation of the German energy system. Core assumption of ENERGY-TRANS was that this transformation is not only a technical but also a societal challenge. The KIT coordinated the Alliance and the duration was from September 2011 until December 2016. More than 80 social and political scientists, psychologists and philosophers, economists, engineers and systems analysts from eight institutions collaborated in the Alliance working together on 17 projects, 2 horizontal tasks and 2 integrative key topics. Besides the University of Stuttgart, ITAS provided the largest research contingent and participated in ENERGY-TRANS with about 20 scientists. Corresponding to the complex tasks of ENERGY-TRANS, the Alliance integrated ITAS' expertise from different disciplines. The highly interdisciplinary research also addressed important issues of the Helmholtz program 'Technology, Innovation and Society' (TIS). In the horizontal task 'Sustainability Monitoring' the objecive was to elaborate a tool for the sustainability assessment of the German energy system and its transition. This comprised the development of a set of indicators and according target values to assess states or development paths for the transformation of the energy system and the identification of already existing or potential future sustainability-related conflicts of goals or interests. The developed Sustainability Indicator System (SIS) is a unique comprehensive tool to assess progress towards a more sustainable energy system and is, thus, useful to support decision-making. It includes several new indicators to assess the interfaces of the system that are lacking in existing indicators sets such as the German monitoring report 'Energy of the Future'. They mainly address the interface between technology and society, which goes far beyond particular monetary aspects such as the costs for electricity supply. The SIS can help to reveal and eradicate the blind spots and weaknesses of existing indicator sets and to improve the assessment of issues at the socio-economic-technical interface of the energy system and its transition. Nevertheless, additional research and methodological work is required to improve the SIS mainly with respect to sustainability issues neglected so far.

Table of contents

Abbreviations

BDI	Bundesverband der Deutschen Industrie
CSD	Commission on Sustainable Development
DALY	Disability-Adjusted Life Years
DDT	distance-to-target
DLR	Deutschen Zentrums für Luft- und Raumfahrt
EEA	European Environment Agency
EEG	Erneuerbaren Energien Gesetz
EPO	European Patent Office
GDP	Gross Domestic Product
ICoS	Integrative Concept of Sustainable Development
MDGs	Millenium Development Goals
OECD	Organisation for Economic Co-operation and Development
PV	Photovoltaic
R&D	Research and Development
SAIDI	System Average Interruption Duration Index
SDGs	Sustainable Development Goals
SI	Sustainability Indicator
SIS	Sustainability Indicator System
TIS	Technology, Innovation and Society
UK	United Kingdom
WGBU	Wissenschaftlicher Beirat der Bundesregierung Globale Umweltveränderungen

Abstract

The goal of the energy transition in Germany is to achieve a sustainable supply of energy. Providing advice for decision-makers, either to continue the current transition pathway or to implement strategic adjustments requires a comprehensive decision support tool. The authors have developed a Sustainability Indicator System (SIS) consisting of 45 indicators that are described in detail in 45 factsheets to assess if policy measures implemented so far by the Federal Government are appropriate and sufficient to achieve the energy policy and sustainability targets defined. Given that at least five years of data were available, the assessment was carried out with the distance-to-target method assuming a linear development of the trend line until 2020 and aligned with a traffic light colour code. The results show that about a half of the assessed indicators are rated with a red traffic light. This means that the targets for these indicators will not be achieved by 2020 without substantial modifications of political strategies and measures implemented so far. The SIS enables the identification of the strengths and weaknesses of the energy system and its transition as well as a systematic analysis of interdependencies and conflicts of goals between sustainability aspects and indicators. Based on this orientation knowledge, resilient political strategies for a successful energy transition can be developed.

1 Introduction

The energy system plays a key role to realize a more sustainable development at the global and national level. The Federal Government has determined political targets and adopted measures to transform the energy system into a more sustainable one [1]. The overall objective is to establish a secure, affordable and ecologically compatible energy supply without nuclear power and based on a growing share of renewables and an increasing energy efficiency. The monitoring process 'Energy of the future' was established to ensure the energy transition develops in the desired direction. Thereby indicators are used to take annual stock of the progress made in attaining the quantitative targets of the German Energy Concept [2–5]. The core indicators are referring to well-known characteristics of the energy system, like the share of renewable energies and greenhouse gas emissions. Social aspects such as the fair allocation of benefits and burdens of the energy transition among social groups or the participation of citizens in the transformation process are to a large extent missing. The scientific expert commission accompanying the monitoring process has drawn attention to this deficiency. The experts recommended not to only focus on 'classical' indicators for which statistical series and targets are available but also to take into consideration further aspects referring to affordability, participation and acceptance [6–8]. In the light of the above, a comprehensive indicator system was developed as an analytical tool to assess the sustainability of the German energy system and to support the development of resilient political strategies for a successful energy transition.

Since the idea of sustainable development is common ground in scientific and political contexts, a number of guidelines, frameworks and tools have been developed to assess the sustainability of technologies, processes and systems [9–13]. The latest and most relevant work in this respect are the 17 Sustainable Development Goals (SDGs) defined by the UN [14] including 169 indicators substantiating these goals. The SDGs partly build upon the Millennium Development Goals (MDGs) that have been adopted by the UN in 2000. The SDGs go much further than the MDGs by addressing the reasons for poverty and the universal need for a development that works for all people. Each SDG has specific targets to be achieved over the next 15 years. One target explicitly refers to energy compromising the objectives to ensure universal access to affordable, reliable and modern energy services and to increase substantially the share of renewable energy in the global energy mix. The SDGs were developed and agreed upon by developed and developing countries, whereas transformative action is dedicated primarily to the national level. Here, more differentiated indicators are needed for striking a careful balance between different sustainable development issues. For the definition of additional indicators with relevance to scientific debates and societal and political decision-making, a theoretically well-founded and operable conceptual approach for analyses and assessments is required. The Integrative Concept of Sustainable Development (ICoS) developed within the German Helmholtz Association [15] is such a concept and is used in this work as a methodological framework to derive a coherent system of sustainability indicators.

1.1 The Integrative Concept of Sustainable Development

Since almost 30 years, several approaches to conceptualize sustainable development have been developed and applied such as the three or four pillar model or the pillar-overarching integrative approaches [16,17]. The three-pillar model is dominating political and scientific practice although it is criticised for its lacking theoretical profoundness in justifying sustainable development as overall guiding principle, its systematic neglecting of interdependencies between the pillars, and an insufficient consideration of the postulate of justice and fairness [16,17]. The Integrative Concept of Sustainable Development (ICoS) [15] was developed to overcome these deficits. In contrast to other concepts structured along the economic, ecological and social dimension, it is based upon three constitutive elements of sustainable development, which characterize the key documents of sustainable development like the Brundtland report [18], the Rio Declaration and the Agenda 21 [19]:

(1) Inter- and intra-generational justice, both equally weighted, as theoretical and ethical fundament. Justice is understood as distributional justice with respect to rights and obligations, benefits and burdens.

(2) A global perspective, by addressing key challenges of the global community and developing goals and strategies to achieve them. It also includes a strategic justification to translate globally defined goals into the national and regional context.

(3) An enlightened anthropocentric approach including an obligation of humankind to interact cautiously with nature based on a well-understood self-interest.

These constitutive elements are translated into three general goals and preconditions of sustainable development:

(1) Securing human existence, including basic needs and the capability of human beings to shape their lives on their own.

(2) Maintaining society's productive potential, which consists of natural, man-made, human and knowledge capital.

(3) Preserving society's options for development and action, addressing immaterial needs such as integration in cultural and social contexts, which complement material needs.

These goals are specified by substantial sustainability rules (Table 1.1) forming the core element of the concept. They describe minimum requirements for sustainable development in the sense of a welfare base that need to be assured for all people living in present and future generations.

Table 1.1: Rules of the Integrative Concept of Sustainable Development [15]

Substantial Rules		
Securing human existence	Maintaining society's productive potential	Preserving society's options for development and action
1. Protection of human health 2. Satisfaction of basic needs 3. Autonomous subsistence based on income from own work 4. Just distribution of opportunities to use natural resources 5. Reduction of extreme income and wealth inequality	6. Sustainable use of renewable resources 7. Sustainable use of non-renewable resources 8. Sustainable use of the environment as a sink for waste and emissions 9. Avoidance of technical risks with potentially catastrophic impacts 10. Sustainable development of man-made, human and knowledge capital	11. Equal access for all to information, education and occupation 12. Participation in societal decision-making processes 13. Conservation of cultural heritage and cultural diversity 14. Conservation of the cultural function of nature 15. Conservation of social resources

Conditions to achieve the substantial sustainability	
1. Internalization of external social and ecological costs	6. Society's ability to respond
2. Adequate discounting	7. Society's ability of reflexivity
3. Limitation of public debt	8. Society's capability of government
4. Fair international economic framework conditions	9. Society's ability of self-organization
5. Promotion of international co-operation	10. Balance of power between societal actors

(4) Conditions to achieve the substantial sustainability

A set of rules concerning the conditions to achieve the substantial sustainability was defined addressing the economic, political and institutional framework conditions to fulfil the substantial rules. The internalization of external costs, for example, addresses the approach of implementing the polluter-pays-principle. The discounting rate as well as handling and dealing with public indebtedness are strongly influencing intergenerational justice. The issues of intra-generational and international fairness and solidarity are addressed by the global economic framework conditions and international cooperation.

In the concept, sustainable development is considered as a 'regulatory idea' that inspires political action, and is based on understanding of policy as a polycentric process, involving different actors and institutions. This requires institutional settings to be shaped accordingly and innovations to be developed within a societal dialogue. Thus, rules to overcome particular problems identified were developed [20]. The rule *'Society's ability to respond'* addresses the capability of actors and institutions to distinguish relevant from less relevant problems and to respond to them adequately. The rule *'Society's ability of reflexivity'* aims to consider impacts of acting in one societal sub-system on others in order to reduce or prevent conflicts in advance. The demand to suitably design and implement measures taking into account is addressed by the rule *'Society's capability of government'*. The rule *'Society's ability of self-organization'* is related to the degree to which societal actors are taking responsibility to support sustainable development strategies themselves. This requires avoiding unjustified imbalances of power and of possibilities to articulate and influence processes between actors, an issue that is addressed by the rule *'Balance of power between societal actors'*.

2 Methods

The Sustainability Indicator System was developed based on the Integrative Concept of Sustainable Development (ICoS). The indicator assessment includes a three-step process:

(1) Definition of indicators based on the ICoS and collection and analysis of facts and preparation of data series.

(2) Definition of targets for each indicator for the years 2020, 2030 and 2050.

(3) Calculation of a trendline and assessment of the projected values by the distance-to-target method.

2.1 Definition of indicators

The definition of sustainability indicators based on the ICoS proceeds via two steps. First, relevance decisions have to be taken with respect to the sustainability rules. Then, the relevant rules are contextualized by indicators [21]. Indicators are the most common and popular tool to measure progress towards sustainable development and for any sustainability analysis [22–24]. Besides, they are useful to communicate ideas, thoughts and values and can lead to better decisions and actions, which are more effective by simplifying, clarifying and making information available to policy makers. Additionally, they provide an early warning to prevent economic, social and environmental setbacks [23]. Taking this into account, the development of indicators faces several challenges and requirements [25]. One issue refers to the suitable number of indicators, allowing for both an appropriate substantiation of goals and manageability of analyses and communicability of results. Another important aspect concerns the appropriate combination of different indicator types, including both context-adapted single or socio-economically differentiated, objective indicators and subjective indicators such as contentment, fears and expectations. This is why, in selecting indicators, it is important to have a sound combination of a science-based 'top down' and a stakeholder-based 'bottom up'-approach.

The set of substantial sustainability rules and rules concerning the conditions to achieve the substantial sustainability, provide basic orientation for development, as well as criteria to assess different states or development paths. They are, a priori, universally valid and unweighted. Thus, conflicts between rules cannot be solved by a hierarchical decision through prioritizing of specific rules. Nevertheless, priorities and weightings, i. e. relevance considerations, are possible and even necessary at the level of particular thematic, regional or other contexts. With this sophisticated architecture and the elements outlined above, the Integrative Concept of Sustainable Development is considered to be a multi-level concept, theoretically well-founded, clearly defined and non-arbitrary that provides a good fundament for setting-up a theory of sustainable development [16]. For assessing the sustainability of the German energy system, not all rules have a clearly definable relation to the subject of investigation. This is true for example for the rule *'Conservation of cultural heritage and cultural diversity'* focusing on cultural treasures such as heritage-listed buildings or precious historical items saved for present and future generations. Similar considerations apply to the rule *'Fair international economic framework conditions'*. This rule is only weakly connected to the German energy system since it focuses on global economic and political framework conditions that are created by supra-national institutions and can neither be influenced nor controlled by the German energy system. Beyond the lack of relevance, some sustainability rules were not concretised due to principle problems of getting significant and reliable information to define feasible indicators. This applies to the rule *'Limitation of public debt'*, because no information was availbale how the energy transition is contributing directly or indirectly to the increase of public debt over time. This applies also for the rule *'Adequate discounting'*. This rule has not been included in the sustainability assessment due to the difficulty to raise valid and representative data for the discounting practice in public and private investment decisions.

The development of the indicator system comprises three steps:

(1) Translation of sustainability rules into indicators based on a literature review and the assignment of indicators to the relevant rules.

(2) Reduction of the number of indicators by applying the criteria comprehensiveness, availability of data, and possibility to determine targets.

(3) Adjustment of the indicator system in response to the feedback of experts (written comments, workshop evaluation) and the results of stakeholder interviews.

In Table 2.1, the main literature used in the comprehensive and in-depth review process is listed. Especially German literature was considered to conduct the queries as efficient as possible. Besides, the intention of the work was to emphasize deficits of existing indicator sets used in Germany. For that reason, the selected indicators are a priori applicable only for the assessment of the German energy system.

Table 2.1: Main literature used for the selection of sustainability indicators

Author / Editor	Report
Agentur für Erneuerbare Energien (AEE) 2015 [26]	Bundesländer mit neuer Energie – Jahresreport Föderal Erneuerbar 2014/2015
Agentur für Erneuerbare Energien (AEE), Deutsches Institut für Wirtschaftsforschung (DIW), Zentrum für Sonnenenergie- und Wasserstoffforschung Baden-Württemberg (ZWS) 2014 [27]	Vergleich der Bundesländer: Analyse der Erfolgsfaktoren für den Ausbau der Erneuerbaren Energien 2014 – Indikatoren und Ranking. Endbericht
Bundesverband der Deutschen Industrie e.V. 2014 [28]	Energiewende-Navigator 2014
Bundesministerium für Wirtschaft und Energie 2016 [5]	Fünfter Monitoring-Bericht 'Energie der Zukunft'
Ecoplan, Factor 2001 [29]	Nachhaltigkeit: Kriterien und Indikatoren für den Energiebereich. Endbericht für das Bundesamt für Energie (CH)
Compiling and Refining Environmental and Economic Account (CREEA) 2014 [30]	Compiling and refining environmental and economic accounts. Ergebnisberichte des EU-Projekts
Statistisches Bundesamt - Destatis 2014 [31]	Nachhaltige Entwicklung in Deutschland – Indikatorenbericht 2014
Statistisches Bundesamt - Destatis 2016 [32]	Umweltökonomische Gesamtrechnungen – Nachhaltige Entwicklung in Deutschland – Indikatoren zu Umwelt und Ökonomie – 2016
Expertenkommission zum Monitoring-Prozess 'Energie der Zukunft' 2016 [8]	Stellungnahme zum fünften Monitoring-Bericht der Bundesregierung für das Berichtsjahr 2015
International Atomic Energy Agency (IAEA), UN Department for Economic and Social Affairs, International Energy Agency, Eurostat, European Environment Agency 2005 [33]	Energy indicators for Sustainable development: Guidelines and methodologies
IASS 2013 [34]	Beiträge zur sozialen Bilanzierung der Energiewende
A.T. Kearney, WirtschaftsWoche 2012 [35]	Energiewende-Index
Ministerium für Umwelt, Klima und Energiewirtschaft Baden-Württemberg 2015 [36]	Monitoring der Energiewende in Baden-Württemberg – Schwerpunkt Versorgungssicherheit und Effizienztrends – Statusbericht 2015
Zentrum für Europäische Wirtschaftsforschung (ZEW) 2012 [37]	Indikatoren für die Energiepolitische Zielerreichung

The literature review resulted in a very large number of indicators at different scales and levels. Since an indicator system, which consists of a very large number of indicators, is difficult to handle regarding reporting requirements, trade-offs between indicators and consideration processes in politics, the number of indicators was reduced to a number below 50. Politically relevant work on sustainability indicators was used as reference to determine this number. The UN Commission on Sustainable Development (CSD), for example, uses 59 core indicators [14] and based on them, Eurostat and the European Environment Agency (EEA) apply 37 indicators, however, only covering environmental issues [27]. A smaller number of indicators is frequently asked for by policy-makers but not feasible from the scientific point of view due to the complex and comprehensive nature of sustainable development. The chosen indicator system is entitled to represent this rather sophisticated reality – the state of a society, its social, economic and ecological connections, and their development and targets. In order to reduce the number of indicators the comprehensiveness of the indicators as well as their suitability to address one or more sustainability rules were crucial for the selection of key indicators. Besides the key indicators were selected in a way that all relevant rules of the Integrative Sustainability Concept were covered. Another selection criterion was the availability of relevant and free accessible data, and the possibility to determine political or desirable targets [26]. Finally, we discussed the reduced set of sustainability indicators with experts in a workshop (Figure 2.1). In order to also include the opinion of stakeholders on the proposed set of sustainability indicators, we carried out interviews in the region of South-West Thuringia.

Figure 2.1: Approach of designing the Sustainability Indicator System

2.2 Definition of targets for the years 2020, 2030 and 2050

Since a distance-to-target (DTT) approach was applied for the indicator-based assessment of the energy system and its transition, targets obviously have a key function. The targets defined are important reference lines for indicator values to be compared with. Strategically, they should allow for higher planning reliability of actors, in particular if targets are designed stepwise over time, and help decision-makers to design political measures. From the DTT approach, the necessity aroused to define targets for all indicators in the SIS. However, not for all of the defined indicators political justified and binding targets were available, since the indicators selected to cover the socio-technical interface of the energy system are rather new. Thus, political discussions and processes of target setting in these cases are still ongoing or even missing. Therefore, we have carried out a comprehensive and profound review of documents from policy consulting institutions, such as the German Advisory Council on Global Change, science, NGOs, unions and other stakeholders, and the media as well as the target agreements of other comparable countries to identify and adopt appropriate proposals for binding or non-binding targets. The objective of that wide-ranging investigation was to define target values for all indicators of the SIS in a comprehensive and reliable way.

As a result, the work comprises a mixture of set, proposed or desirable targets with different degree of justification by politics and society: Some of them have been derived from policy-based targets in 2020, both binding and non-binding, some were adopted from political targets or good examples in other countries, some from policy consulting institutions, some from science and other targets have been abstracted from public debates. As described above, the targets were determined based on these different sources, for the years 2020, 2030 and 2050. Primarily, political targets were adopted if available, either at the national or international scale. To give examples: For the indicators 'Primary energy use', 'Energy-related greenhouse gas emissions' and 'Number of electric vehicles' the political targets defined by the German government were used. For the indicator 'Energy-related emissions of mercury', the targets were taken from a United Nations protocol. In cases where targets exist only for 2050, the authors determined according values for 2020 and 2030, mainly based on a linear extrapolation.

Secondly, targets were adopted or derived from scientific or societal debates as e. g. for the indicator 'Relation of technician salary to manager salary in the big electricity suppliers' that refers to the Swiss debate, and for the indicator 'Area under cultivation of energy crops', following recommendations of the German Advisory Council on Global Change. In addition, a cross-border look at other countries' best practices provided a source to derive targets. One example here is the indicator 'Federal expenditures for energy research'. The research spending in Germany in relation to its GDP and the research spending of the country with the highest value in this category (South Korea) are used as reference point for future expenditures.

For those indicators where no targets were available or discussed, either in politics or in science, conclusion by analogy was chosen to define targets. To give an example: for the indicator 'Final energy consumption of private households per capita', the political target for national primary energy use was adopted. A similar procedure was applied for the indicator 'Number of university graduates in energy sciences', assuming that this indicator develops proportionally to the volume of investments in Germany given in the DLR-Report [9], which provided the key basis for all model-based analyses in the project. For the indicator 'Number of start-ups in renewable energy and energy efficiency sector', targets were defined in accordance with the indicators 'Number of German patents in the field of renewable energy and energy efficiency' and 'Federal expenditures for energy research'.

For those indicators where no political targets existed at all, either at the national or international scale, the authors adopted or derived targets from scientific advisory bodies or suggestions that arose in societal debates. The target for the indicator 'Area under cultivation of energy crops' was definded based on recommendations of the German Advisory Council on Global Change. The Swiss debate on just wages was used as a blue print to define a target for the indicator 'Relation of technician salary to manager salary in the big electricity suppliers'. The initiative '1:12-Initiative for just wages' in Switzerland want to anchor that no manager is supposed to earn more than one month's worst paid employee per year in the Swiss constitution.

2.3 Indicator assessment based on the distance-to-target approach

After the definition of targets for 2020, 2030, and 2050, the performance of the sustainability indicators was assessed based on a combined linear extrapolation and distance-to-target (DDT) approach taken from the

German monitoring report 'Energy of the future'. Accordingly, a linear projection of the performance trend for each indicator was calculated based on the previous five years for which data were available, assuming that this trend will continue in a linear way until 2020. Then, this extrapolated trend was compared to the targets for 2020, in order to assess to which degree the target will be met within the framework of the existing energy policy. The near-term target 2020 was chosen because here a linear projection is regarded as feasible since it can be assumed that the framework conditions influencing the energy system will remain relatively constant within this short time period, and that effects of measures previously implemented will support the trend until 2020. For the period until 2050, however, it can be expected that due to the unpredictable nature of the complex and dynamic energy system, as well as changing political and institutional framework conditions, indicator performance trends will change accordingly and, thus, extrapolation is not a valid methodology any more. The traffic light symbol was used to visualize the assessment results (Figure 2.2).

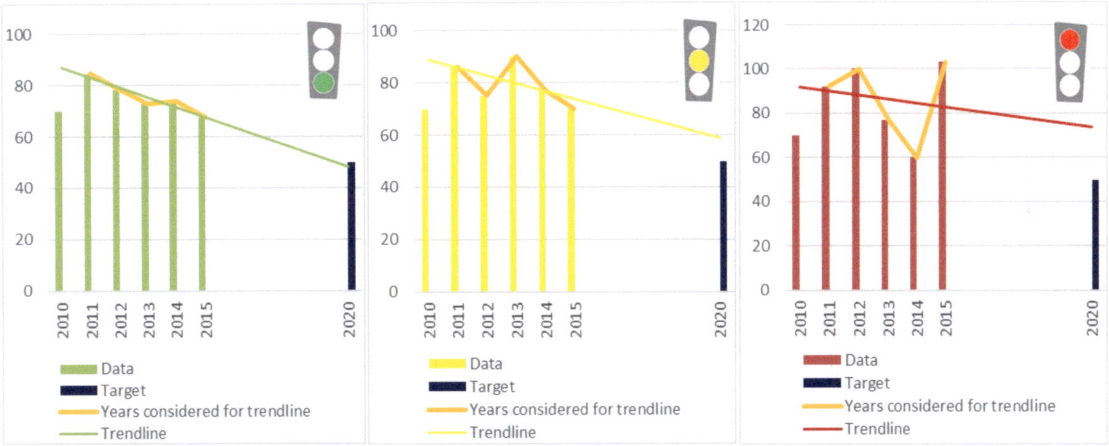

Figure 2.2: Approach to assess the sustainability indicators applying distance-to-target considerations

The assessment includes the following steps:

(1) Defining a 'reference value' by calculating the average value of the previous 5 years with data

(2) Calculating the necessary change (relation between 'reference value' and 'target value')

(3) Calculation of a 'projected value' for 2020 by extrapolating the trendline, covering the past 5 years with data until 2020

(4) Calculation expected change (relation between 'reference value' and 'projected value')

(5) Calculation of the relation (in %) between the necessary change and the expected change according to the following formula:

$$\left(1-\frac{1-\frac{PV_{2020}}{AV_5}}{1-\frac{TV_{2020}}{AV_5}}\right)*100\% \tag{2.1}$$

PV2020: Projected value for 2020; TV2020: Target value of 2020; AV5: Average value of past 5 years with available data

Based on the relation of the projected value for 2020 (expected change) to the target for 2020 (necessary change), the deviation measures if the target will be met or missed. The deviation to the target is expressed as the distance to the target (see Figure 2.2).

If the deviation from the target 2020 is more than 40 % or the calculated trend goes in the wrong direction, a red traffic ligtht is assigned, meaning that there is a high probability that the target will be missed if business and politics continue as usual and that changes of political actions are urgently required in order to still reach the target in 2020.

A yellow traffic light should be understood in the sense, that the deviation from the target 2020 is more than 10 and up to 40 % and that the target will probably not be met if policy remains unchanged. In this case, a moderate change of the strategies and measures is recommended to achieve the target.

A green traffic light means that the projected value for 2020 will meet or exceed the target value in 2020, or the deviation is up to 10 %.

A white traffic light is assigned, if no DDT approach can be carried out due to lacking data series.

3 Results

The developed Sustainability Indicator System (SIS) is presented in four sections structured along the three general sustainability goals 'Securing human existence', 'Maintaining society's productive potential' and 'Preserving society's options for development and action' as well as the goals concerning conditions to achieve the substantial sustainability (Tables 3.1, 3.2, 3.3, 3.4 and 3.6). The indicators are described in detail in 45 factsheets in the annex. Before presenting the results in detail, a few characteristics regarding the SIS shall be highlighted. First, the SIS does not represent the entire sustainability rules due to relevance considerations (see chapter 2.1). Another important issue to be considered is that some rules are addressed mainly by so-called classical indicators that are well-known from political agenda settings. For other rules, however, it was necessary to develop and define 'new' and partly provisional indicators, for example for the rule *'Participation in societal decision making processes'*. In a few cases, an appropriate indicator could not even be proposed. To give an example: The energy transition and respectively the broader implementaion of renewable energy technologies have various impacts on biodiversity. The different direct and indirect impacts, however, cannot be merged into a meaningful indicator, which can be assessed easily. Thus, the authors have deliberately dispensed to propose an indicator on biodiversity.Since the SIS compromises indicators which have an impact on biodiversity, such as land use, an indirect assessment is possible.

Looking at the SIS it becomes apparent that some rules, are put in practice with more than one indicator. Since no standard exists determining that one rule has to be addressed by only one indicator, it is possible that the number of indicators addressing the rules can vary widely. For some rules, it was regarded as sufficient to define one indicator. For other rules, however, more than one indicator was needed to address the different facets of the rule adequately and sufficiently. To give an example: The rule *'Sustainable use of non-renewable resources'* is reflected by different political areas of activities and targets, such as the energy consumption of households, transport and industry as well as the modal split. Therefore, this rule is addressed through eight indicators (Indicator No. 13 to No. 20). It could be argued that any rule that has more than one indicator gets a relatively higher weight in the evaluation system. This, however, is not the case because within the evaluation process all rules are defined as being a priori equally weighted. Decisions about the relevance and relative importance of rules and indicators can only be taken at the level of particular thematic, regional or other contexts by politics and society.

3.1 Assessment of indicators to secure human existence

The sustainability goal to secure human existence is defined by five rules and addressed by nine indicators (Table 3.1). The first three indicators address the sustainability rule *'Protection of human health'* and concern the energy-related emissions of particulate matter and the emissions of cadmium and mercury; all of them are harmful to human health as they cause a wide range of serious health problems (see Factsheets No. 1 to 3 in the annex). Alternatively, human health could be addressed by the concept DALY (Disability-Adjusted Life Years) which calculates the life years lost due to work-related diseases and lethal and non-lethal occupational accidents in the energy sector. Another option is to calculate the fatality rates of energy technologies based on the energy-related severe accident database [38,39]. Both concepts are based on elusive assumptions and system boundaries. As neither direct health impacts of technologies can sufficiently be assigned to the energy sector, the most important energy-related emissions have been selected as indicators to address the issue of human health.

Table 3.1: Indicators, targets and assessment of the sustainability goal 'Securing human existence'

Sustainability rule	Sustainability indicator	Targets 2020	Targets 2030	Targets 2050	Target references	Assess-ment
Protection of human health	1. Energy-related emissions of particulate matter (kt)	67.4	60.1	45.6	Target 2020 based on the amended protocol of Gothenburg, assumptions for 2030 and 2050.	green
	2. Energy-related emissions of cadmium (t)	3	2.7	2	Target 2020 developed by analogy based on the amended protocol of Gothenburg, assumptions for 2030 and 2050.	yellow
	3. Energy-related emissions of mercury (t)	6.5	5.8	4.4		green
Satisfaction of basic needs	4. Energy import dependency (%)	69	58	43	Targets taken from [40].	red
	5. Monthly energy expenditures of households with a monthly net income less than 1,300 Euro (€/month)	139	142	147	Targets developed by analogy based on [41] and assumptions.	green
	6. SAIDI of electricity (min)	12.5	10	10	Targets 2030 and 2050 developed by analogy based on [42].	green
Autonomous subsistence based on income from own work	7. Relation of employees in the renewable energy sector to total employees (%)	0.94	0.93	1.19	Targets developed by analogy based on [40], [43] and assumptions.	red
Just distribution of opportunities to use natural resources	8. Final energy consumption of private households per capita (GJ/capita)	29.3	24.7	17.6	Targets developed by analogy based on [44].	green
Reduction of extreme income and wealth inequalities	9. Relation of technician salary to manager salary at the big electricity suppliers	1:12	1:12	1:12	Targets developed by analogy based on [45].	red

The energy adresses the rule *'Satisfaction of basic needs'* and has a physical and a technical, but also an economic dimension, which are addressed by three indicators. The indicator 'Energy import dependecy' characterizes one aspect of the physical vulnerability of energy supply, by addressing Germany's dependency on imports (see Factsheet No. 4). Regarding the target of this indicator, it has to be taken into account that a high technology country like Germany can alleviate but not completely reduce its dependency from imports. To address the technical security of electricity supply, the indicator 'System average interruption duration index' (SAIDI) was selected (see Factsheet No. 6). The SAIDI shows the average supply interruption from the electricity grid, measured in minutes per year per connected consumer.

The affordability of energy supply was addressed by focusing on the energy expenditures for low-income households. This social group is supposed to suffer from 'energy poverty'. This means that they are not able to adequately heat their homes or use other energy services at affordable costs due to rising energy prices, low income and poor energy efficiency of heating systems or other devices [41,46]. Experiences in Germany show that the energy transition leads to growing energy expenditures of households because the costs to increase the share of renewable energies are allocated to customers through the Erneuerbaren Energien Gesetz (EEG) (Renewable Energy Law) shared contributions. This allocation system has been discussed controversially. The impact of this financial burden on the energy expenditures of low-income households has been associated with terms such as 'energy poverty' or 'fuel poverty'. However, there is little agreement even on the problem definition and the measuring method. Moreover, evidence exists that the assessment if and to which extent 'fuel poverty' exists strongly depends on the measuring method used [47]. Hence, the authors refrained from an evaluation of data without mathematical methods carried out in [48], and propose to determine 'essential expenditures' of low-income household for an adequate energy supply for electricity and heating, according to the recommendation of [48]. The statistically raised data about energy expenditures of low-income households should be compared to these 'essential expenditures'. Not surprisingly, these values have not been determined (even not discussed) in Germany or other countries for different household types, since this is a highly normative decision, hardly justifiable in 'objective' terms. In fact, these expenditures are raised and used to date only within the English Household Survey and were used in the model BREDEM to investigate 'energy poverty' in the United Kingdom (UK). Besides the lack of appropriate poverty targets available from other countries, we chose the target from the UK, because the climatic and economic conditions in the UK are similar to those in Germany.

Beyond the fact that this approach is suitable in general, but not operable to date, the authors propose to refer on the indicator 'Monthly energy expenditures of households with a monthly net income less than 1,300 Euro' as a first approach to monitor if the energy transition leads to undesirable additional financial burden (see Factsheet No. 5). If this might be associated with the term 'energy poverty' remains open to discussion. The monthly net income of households is categorized according to the German Federal Statistical Office and calculated by subtracting income and wage taxes, church tax, and the solidarity surcharge as well as the mandatory social security contributions from gross household income consisting of the total income of the household from employment, property, public and private transfers, and subletting. Data for the monthly energy expenses from 2002 to 2012 for the income class below 1,300 € have been taken from [49]. They include electricity, fees, fuel costs for heating and taxes or levies on heating plants. To derive a data series of five years, data for 2013 have been calculated from information given in [50] and are the weighted average of the income classes below 500 € (2.6 % of this household group), 500 to 900 € (39.5 % of households), and 900 to 1,300 € (57.9 % of households). Data for 2014 and 2015 are taken from [51,52].

In principle, the target for this indicator would have to be adjusted over time considering the development of the income of the group concerned, the development of energy prices and the inflation rate. Since these values are not known, no prediction was made by the authors. Hence, the authors used research results on energy poverty from the UK, where most research on this issue is carried out in the EU. According to [53], in the UK the expenditure of low-income households on heating should not exceed 10 % of their income. A higher percentage would indicate 'fuel poverty'. Despite the critical view of [41] on the data from [53], the authors decided to use this percentage to determine the target, simply because no other valid data were available to define a 'German standard'. On average, German households spend 70 % of their energy expenditure on heating and 30 % on electricity [54]. By weighting these two values, in Germany the expenditures for heating and electricity should not exceed 15 % of the net disposable household income of low-income households. Households in the category 'net income below 1,300 €' had on average a net income of 901 € in 2011 [51] and 916 € in 2015 [52]. Based on these data, values of expenditures for heating and electricity of 135 € in 2011 and 137 € in 2015 were derived.

The data for the period 2011 to 2015 show that households with a net income below 1,300 € spend on average 89 € per month for energy use. Based on the data for the past five years, values for the net income for 2020, 2030 and 2050 have been assessed. For the target values, 15 % of these net income values have been assumed corresponding to 139 € in 2020, 142 € in 2030 and 147 € in 2050. Since the trendline shows a decreasing monthly expenditure not reaching the maximum target value for 2020, a green traffic light was assigned to this indicator. Despite the green traffic light, however, there might be households who suffer from 'energy poverty' because their income is below the average of all households with incomes below 1,300 Euro, which was used as database here.

Against this background, the authors recommend to replace the proposed indicator in the future by a more sophisticated indicator which relates the actual expenditures for electricity and heat of low-income households to the expenditures required by these households for an essential provision with these types of energy. These so-called 'essential' expenditures still need to be defined for vulnerable household types [48].

Autonomous self-subsistence is an important requirement for sustainable development as it refers to the possibility of human beings to secure their livelihood by a freely chosen occupation. The energy sector is an important employer and the continuing growth of jobs in the renewable energy sector is significant. This increase is being driven by declining renewable energy technology costs and enabling policy frameworks. The indicator 'Relation of employees in the renewable energy sector to total employees' was selected although we were fully aware that jobs in this new sector will reduce employment in the 'old' fossil fuel based energy sector (see Factsheet No. 7). Besides, employment in other sectors could at first decline due to increasing energy costs caused by a higher share of expensive renewable energy. Then, employment and also prices could decrease in the future if the new energy sector turns out to be very efficient over time. An increase in the efficiency of electricity production is linked with a decrease in labour costs that could improve the overall employment rate. However, an increase in employment is mainly based on plant construction, operation and maintenance. In view of these considerations, the defined indicator is regarded as provisional indicator that need to be improved or even replaced by a more comprehensive one including all direct and indirect employment effects of the energy transition if data are available.

The provisional indicator 'Relation of employees in the renewable energy sector to total number of employees' includes the employment due to domestic production for domestic use and for exported renewable energy compounds. It comprises employees responsible for maintenance and operation of renewable energy plants, but excludes employment due to the production in other countries, e.g. the production of photovoltaic modules in China, since the sustainability analysis is focusing on Germany. A decline of employees in the conventional energy sector and other sectors as direct consequence of the energy transition is not taken into account, also higher energy costs resulting from subsidies for renewable energies (indirect effects) due to the lack of reliable data series. Higher energy prices can have an impact on consumption and production and thus on total employment. This indicator shows continuously increasing values from 2007 to 2012, mainly because the number of employees in the renewable energy sector steadily increased from 277,300 in 2007 to 399,800 in 2012. Then the number decreased to 371,400 in 2013, to 355,000 in 2014 and to 330,000 in 2015 [5,55,56]. The share of employees in 2007 to 2015 was calculated based on these data and data of total employees given in [43].

The number of employees in the renewable energy sector mainly depends on the volume of investments into this sector in Germany, the export of renewable energy technologies, and the maintenance and operation intensity of renewable energy plants. Model-based information on the volume of investments in Germany until 2050 is given in [40]. Data on future exports and for employees responsible for maintenance and operation of renewable energy plants are not available. Therefore, the authors estimated the number of employees for the years 2020, 2030 and 2050 based on the estimated volume of investments in the field of renewable energy. In 2015, investments in the construction and maintenance of renewable energy plants (not investment in general) amounted to 15 billion euros [5] and the number of employees was 330,000. The yearly volume of future investments has been taken from [40]. It accounts for 18.4 billion euros until 2020, 17.2 billion euros until 2030, 18.7 billion euros until 2040 and 19.9 billion euros until 2050 [40]. Based on these numbers, 416,000 employees for 2020, 387,000 employees for 2030 and 449,000 employees for 2050 were calculated. However, an even larger increase of gross employment from 530,000 to 640,000 people in 2030 would be possible assuming that a global technological leadership of the German industry also leads to a considerable competitive advantage on the growing future world energy market [40].

According to [43], the total number of employees was 41.5 million in 2011 and 43 million in 2015. Starting from the average value of the relation of employees in the renewable energy sector of 0.87 % over the past 5 years (2011-2015), the targets in Table 3.1 were calculated, using the data given in [40] for the renewable energy investments and the total number of employees. The calculated trendline to 2020 shows a decrease of about 34 %, whereas the target recommends an increase of about 8 %. This leads to the assignment of a red traffic light for this indicator.

The rule *'Just distribution of opportunities to use natural resources'* implicates the issue of fair allocation of chances, responsibilities and burdens among all people with respect to natural resource use. The fewer natural resources and environment absorption capacities are used in Germany, the more are available for people in other countries. The indicator 'Final energy consumption of private households per capita' is addressing this issue (see Factsheet No. 8). The comparison at the global scale reveals that, for example, the household electricity consumption in Germany per year is higher than the world average and 25 % above the electricity consumption in Italy [57].

The sustainability rule *'Reduction of extreme income and wealth inequalities'* should be matched to combat poverty and social marginalization. Both are neither related to distinct economic sectors nor to technologies, but rather caused by social and tax regulations. However, huge disparities between the salaries of employees can consolidate or further increase inequalities. Therefore, the relation between technician salary and manager salary within the big energy suppliers has been defined as indicator. This relation has significantly worsened in the last years and amounts to 1:110 in 2015 (see Factsheet No. 9).

3.2 Assessment of indicators to maintain society's productive potential

The sustainability goal to maintain society's productive potential is defined by five rules that are translated into 22 indicators (Table 3.2).

The rule *'Sustainable use of renewable resources'* is addressed by the classical indicator 'Share of renewable energy in gross final consumption of energy' (see Factsheet No. 10) and the new indicator 'Area under cultivation of energy crops' (see Factsheet No. 11) which was defined due to the evidence, that energy production with renewables signifcantly increases the demand for land. For the cultivation of energy crops, agricultural land is required. Land, however, is a finite and increasingly scarce resource. This leads to competition or even conflicts with other land uses, such as for food, feed and fibre production. Land is also needed for the installation of renewable energy plants, such as biogas plants, open space PV systems or wind energy plants, as well as power transmission lines. Compared to the land use requirements for conventional energy production with fossil fuels, for example for the installation of power plants or mining of brown coal, the energy transition towards renewable sources is associated with a higher land use. The type and extent of land use is quite different among the energy technologies and therefore cannot be summed up. In Germany, 12.3 % of the agricultural area is used for energy cropping [58] indicating a strong impact on land use. There is a controversial debate on land use for energy production in the face of world hunger, widely known as the food versus fuel debate, and evidence on climate impacts and other adverse environmental and social impacts of energy crops [59].

Land use data for the cultivation of energy crops are given in [58]. However, the different kinds of land use listed in [7] should not be summed up, because they are associated with different sustainability-related impacts. In addition, parts of the land occupied by energy production can still be used for other purposes or can be re-cultivated after the energy production phase. Therefore, the authors have decided to take into consideration only land use for the cultivation of energy crops. The cultivation of energy crops requires agricultural land and, therefore, will further lead to an increase of competition for land [18]. This growing demand can be satisfied by extending cropland and pastures into new areas, thereby replacing natural ecosystems, and/or by improving productivity of existing cultivated land through an increasing or more efficient use of inputs, improvement of agronomic practices and crop varieties, etc. Both options have negative environmental impacts, for example on the conservation of biodiversity. The import of biomass for food, feed, fuels and industrial applications is regarded as an unsustainable strategy to reduce land use conflicts, because this will only shift such conflicts to other countries. The land footprint abroad to satisfy the German (bio)energy demand has not been taken into account here, because the system boundaries defined for the SIS only comprise processes located in Germany, and due to lacking valid data. The trend calculated based on data for the past 5 years (2011–2015) shows an increase for this indicator of about 11 % by 2020 compared to the average value for 2011 to 2015.

According to [60], it is necessary to determine limits for the area dedicated to energy cropping in order to minimize land use conflicts. The authors derived these limits from two general principles based on the Sustainable Development model. First, to reach the SDG No. 2 (stop hunger and all forms of malnutrition by 2030), the production of food must be given priority over the production of renewable energy sources or the use for terrestrial CO_2 storage. Thus, it is hardly justifiable to convert arable land from food production to energy cropping. Second, land use for energy crops should not jeopardize the nature conservation target determined by the German Advisory Council on Global Change (WGBU). The WGBU has proposed that 10-20 % of the total land area should be reserved for nature conservation to protect, restore and promote a sustainable use of terrestrial ecosystems, and to minimize biodiversity loss. Since worldwide only 8.8 % of total land area are designated as protected areas (category I-VI areas), the conversion of natural ecosystems to land cultivated for energy crops has to be rejected as a matter of principle. As a global benchmark, the WBGU recommends to allocate not more than 3 % of the terrestrial area to energy cropping to avoid conflicts with nature conservation. Considerations of particular regional conditions and possibilities are indispensable to translate this global target into the national scale. As recommended in [60], a maximum of 10 % of arable land and 10 % of pasture land should be used for the cultivation of energy crops in Europe. According to [60], these two percentages correspond to an area of 22 million ha or 4.5 % of the land area available for the cultivation of energy crops in the European Union due to the decline in agricultural land.

This target is used for calculating the potential area in relation to the total land area of 34.867 million ha in Germany [61]. In doing so, the calculated target to be achieved by the year 2050 is about 1.57 million ha used for energy crops as a maximum. The targets for the years 2020 and 2030 were derived by interpolation from the target for 2050. Based on the average value of 2.13 million ha for energy crops over the years 2011 to 2015 and the target for 2050, the following targets were derived by linear interpolation: for the year 2020 a target of 2.0 million ha (5.6 % of the land area of Germany), and for the year 2030 a target of 1.9 million ha (5.4 % of the land area of Germany). In order to achieve the target of 2.0 million ha for 2020, a reduction by 4.7 % of the energy crops area compared to the mean value of 2.13 million ha for the years 2011 to 2015 is required. Since the trendline shows a further increase in the area under cultivation of energy crops, this indicator is aligned with a red traffic light.

The indicator 'Unused renewable electricity due to management measures' was defined, because the installed capacities to produce renewable energy have to be used in a more efficient way and temporarily reduction of production to avoid an overload of the grid and blackouts have to be reduced (see Factsheet No. 12).

Table 3.2: Indicators, targets and assessment of the sustainability goal 'Maintaining society's productive potential' - part 1

Sustainability rule	**Sustainability indicator**	**Targets 2020**	**2030**	**2050**	**Target references**	**Assess-ment**
Sustainable use of renewable resources	10. Share of renewable energy in gross final consumption of energy (%)	23	36	60	Targets taken from [40].	🔴
	11. Area under cultivation of energy crops (mio. ha)	2	1.9	1.6	Targets developed by analogy based on [60].	🔴
	12. Unused renewable electricity due to management measures (GWh)	4,047	2,698	0	Target 2050 based on assumptions, linear extrapolation for 2030 and 2020.	🔴
Sustainable use of non-renewable resources	13. Use of primary energy (PJ/a)	11,504	10,066	7,190	Targets taken from [44].	🔴
	14. Specific final energy consumption of households for heating (temperature-corrected) (MJ/m^2)	435	367	230	Targets taken from [40].	🔴
	15. Final energy consumption in the transport sector (PJ)	2,337	1,973	1,521		🔴
	16. Modal split in the transport sector (%)	20	20	20	Targets developed by analogy based on [40] and own assumption	🟡
	17. Number of electric vehicles (mio.)	1	6	22	Target 2020: political goal of the Government, target 2030 based on [44], target 2050 taken from [40].	🔴
	18. Final energy productivity of the German economy (€/GJ)	366	482	743	Targets taken from [40].	🔴
	19. Final energy productivity of the industry (€/GJ)	306	403	621		🟡
	20. Final energy productivity of trade, commerce and services (€/GJ)	1,602	2,111	3,251		🔴
Sustainable use of the environment as a sink for waste and emissions	21. Energy-related greenhouse gas emissions (mio. t of CO_2 eq.)	622	467	207	Targets derived from political goals of the Federal Government.	🔴
	22. Energy related emissions of acid-forming gases (mio. t of SO_2 eq.)	0.93	0.85	0.69	Targets developed by analogy based on the amendment of the Gothenburg Protocol (see [62]) and own assumptions.	🟢
	23. Energy-related hazardous solid wastes (t)	789,223	526,148	0	Target 2050 based on assumptions, linear extrapolation for 2030 and 2020.	⚪

The rule *'Sustainable use of non-renewable resources'* is substantiated by eight indicators. In addition to the classical indicator 'Use of primary energy' (Factsheet No. 13), the indicator 'Specific final energy consumption of house-holds for heating (temperature-corrected)' was selected, because the existing building stock is a major energy consumer in Germany, aside from the mobility sector for which the indicator 'Final energy consumption in the transport sector' was selected (see Factsheet No. 14 and No. 15). The indicator 'Modal split in the transport sector' measures sustainable transport particularly for non-motorized (cycling and walking) and public transport (Factsheet No. 16). The indicator 'Number of electric vehicles' was defined since electric cars address the sustianble use of non-renwable resources by using electricity from renewables instead of fossil fuels (Factsheet No. 17). The political target of the Federal Government to rise the number of electric vehicles to six millions until 2030 was adopted [1]. Three indicators address the final energy productivity in total, the industry sector and the sector trade, commerce and services (Factsheets No. 18 to 20).

The rule *'Sustainable use of the environment as a sink for waste and emissions'* (see Table 3.2) is substantiated by three indicators referring to energy-related greenhouse gas emissions, emissions of acid forming gases, and hazardous solid wastes because these are the main emissions generated by the energy system (see Factsheets No. 21, 22 and 23). In 2015, the German energy sector was responsible for about 80 % of the greenhouse gas emissions [63] and 43 % of the acid-forming emissions [64].

The rule to avoid technical risks with potentially catastrophic impacts (Table 3.3) is addressed by the indicator 'Amount of high-level radioactive waste which has not been transferred to a save final disposal site'.

Table 3.3: Indicators, targets and assessment of the sustainability goal 'Maintaining society's productive potential' – part 2

Sustainability rule	Sustainability indicator	Targets			Target references	Assess-ment
		2020	2030	2050		
Avoidance of technical risks with potentially catastrophic impacts	24. Amount of high-level radioactive waste which has not been transferred to a safe final disposal place (t HM)			0	Target 2050 based on assumptions.	
Sustainable development of man-made, human and knowledge capital	25. Installed capacity of renewable energy power plants (GW_p)	116	144	169	Targets taken from [40].	
	26. Number of university graduates in the field of energy sciences	2,702	2,516	2,919	Targets developed by analogy based on [40] and assumptions.	
	27. Federal expenditures for energy research (mio. €)	1,212	1,365	1,670		
	28. Number of German patents in the field of renewable energy and energy efficiency	2,580	2,874	3,459		
	29. Number of start-ups in the renewable energy and energy efficiency sector	18,288	20,363	24,515		
	30. Added value creation from the renewable energy sector (billion €)	24.6	29.4	36.4	Targets developed by analogy based on [65], [66] and assumptions.	
	31. Added value creation from energy efficiency measures in households (billion €)	28	35	42	Targets developed by analogy based on [40] and assumptions.	

Radioactive waste, especially spent fuel and waste from reprocessing, involve risks and hazards for man and the environment. In the next years, the amount of radioactive waste will increase due to an increase of spent fuels and the lack of a political decision for a final storage site.

The indicator 'Installed capacity of renewable energy power plants' addresses the rule *'Sustainable development of man-made, human and knowledge capital'* (see Factsheet No. 25). The mix of renewables, the grid development and electricity storage capacities as well as the electricity demand determine the capacity needed.

While innovation is widely considered to be an important engine of the energy transition in Germany and a basic prerequisite to the general sustainability goal of 'Maintaining societies' productive potential', measuring innovation is not easy, since knowledge about innovation processes and results is often limited. Different approaches are available and various attempts have been made to measure innovation. For instance, asking experts in their respective fields to identify major innovations can be one method. However, this provides a rather subjective perspective and it is difficult to gain an overall and continuous picture of innovation. Therefore, the authors propose to use more than one indicator to properly assess the energy-related innovation process at different stages on a quantitative basis, encompassing both, the input into the innovation process and its outcome.

The selected indicators are, first of all, 'Number of university graduates in the field of energy sciences' (see Factsheet No. 26) and 'Federal expenditures for energy research' (see Factsheet No. 27). Research and Development (R&D) expenditures are often used as a proxy for innovation or technological progress. Well-trained young people with different graduate degrees are needed to support the energy transition in practice, teaching and research. Besides, it is imperative to have enough research funds available to boost and sustain human knowledge capital in the field of energy, which is why the indicator 'Number of university graduates in the field of energy science' and the indicator 'Federal expenditures for energy research' were selected. In 2015, the number of these graduates has reached 2,464 [67] and the federal research expenditures in the field of energy were summing up in 2014 to 1,076 mio. euro [68].

However, expenditure is an input for R&D rather than an outcome of R&D, which should be innovation. Therefore, the authors additionally propose the indicator 'Number of German patents in the field of renewable energy and energy efficiency' (see Factsheet No. 28), since patent data and statistics on new technologies are increasingly used to measure innovation, using e. g. European Patent Office (EPO) data, which provides long time data series. Although patent data are frequently used as an innovation indicator, their application is discussed controversially due to the constraints that are associated with this approach [69]. The key argument is that not all patents represent innovation, nor are all innovations patented. Besides, there are a small number of highly valuable patents and a large number of patents with little value. Scherer and Harhoff (2000) showed in their survey of German patents in total that about 10 % of the most valuable patents account for more than 80 % of the economic value of all patents [70].

Against this background, the authors decided to select also the indicator 'Start-ups in the renewable energy and energy efficiency sector' (see Factsheet No. 29), since entrepreneurial activity can be seen as an outcome of innovation processes and an initiation of opportunities opening up in the changing energy market. Niche actors, such as start-ups, play an important role in the energy transition process because they can support the implementation of shifts in the sociotechnical landscape [71] and explore, develop or advance innovative products and processes that are required to shape transition [4]. Particularly when it comes to the commercialization of new energy technologies, start-ups may capture entrepreneurial opportunities or provide complementary niche innovations to the current regime players [72,73].

Data on 5,000 business start-ups used to describe and analyse the indicator are derived from [74]. It was classified according to the 'Environmental Goods and Services Sector' framework. Thus, the start-ups could be assigned to eight distinguished sectors of the green economy: climate protection, renewable energies, energy efficiency, emission prevention, recycling economy, resource efficiency, renewable resources, and biodiversity. Only the firms in the renewable energy and energy efficiency sector were considered for this indicator, in order to avoid duplicates, e.g. firms that are active in more than one sector. The numbers of start-ups taken from [74] differ significantly from those presented in [3] (based on [75]). One reason is that the Centre for European Economic Research [75] uses a more conservative method to ascribe start-ups to the renewable energy sector that is based on a keyword search within the company name and description. The Borderstep Institute, however, uses individual Internet-based research to classify the firms within the sample. In general, this indicator has the problem that the data series ends in 2013.

To determine targets for this indicator, it is assumed that the number of start-ups develops in proportion to the number of registered patents in the renewable energy and energy efficiency sector. The number of newly registered patents, in turn, is assumed to depend on expenditures for energy research. The target for energy research expenditure in Germany was assumed to increase from 2.92 % in 2013 to 4.36 % of the GDP in 2050. This corresponds to an increase by the factor 1.49 by the year 2050 compared to 2013. The target was defined by using the OECD-country with the highest value in the category of research spending in relation to the GDP as reference point, which is South Korea with 4.36 % in 2013 [76]. The research spending for the energy sector is assumed to increase also by the factor 1.49 to ensure that the share of energy research in total research spending remains the same. The same factor is applied to define the target for the number of start-ups in 2050 (24,515). The average number of start-ups over the past 5 years for which data were available (16,420) was used as initial value to derive the targets. The targets for the years 2020 and 2030 were interpolated accordingly, resulting in 18,288 start-ups in 2020 and 20,363 in 2030. The trendline calculated based on the past 5 years (2009–2013) shows a decrease in the number of start-ups of approx. 48 % by 2020 compared to the average value over the years 2009 to 2013. Since the target for 2020 is 11 % higher than the average value for the years 2009 to 2013, a red traffic light is assigned for this indicator.

Another two indicators are addressing the creation of sustainable added value by the renewable energy sector (see Factsheet No. 30) and energy efficiency measures in households (see Factsheet No. 31). They are related to the national level and include company profits, taxes, and income from wages, which are generated e.g. through planning and installation of renewable energies or energetic refurbishment.

3.3 Assessment of indicators to preserve society's options for development and action

The sustainability goal to preserve society's options for development and action is defined by four rules that are translated into five indicators addressing the socio-technical interface of the energy system (Table 3.4). The indicators proposed are rather new in the political and scientific debate.

Table 3.4: Indicators, targets and assessment of the sustainability goal 'Preserving society's options for development and action'

Sustainability rule	**Sustainability indicator**	**Targets**			**Target references**	**Assess-ment**
		2020	**2030**	**2050**		
Equal access for all to information, educa-tion, and occupation	32. Gender pay gap in the highest salary group in the energy sector (€/a)	9,754	0	0	Targets 2030 and 2050 based on assumptions.	●
Participation in societal decision making processes	33. Share of regulatory tools in the planning of power trans-mission grids that fulfill regulatory requirements (%)	92	100	100		○
Conservation of the cultural function of nature	34. Share of tourists who perceive energy power tech-nologies as being disruptive in the vacation area (%)	10	7	0	Targets based on as-sumptions.	○
Conservation of social resources	35. Acceptance of renewable energies in the neighbor-hood (%)	71	81	100	Target 2050 based on assumptions, linear extrapolation for 2030 and 2020.	●
	36. Acceptance of grid extension for achieving 100 % renewa-ble energy supply (%)	72	81	100		○

The rule *'Equal access for all to information, education, and occupation'* is addressed by the indicator 'Gender pay gap in the highest salary group in the energy sector' (see Factsheet No. 32). The gender pay gap belongs to the sustainability indicators proposed by the EU [14]. Still, women in Germany aross all sectors and salary levels earn 23 % less on average than their male colleagues [77]. In an EU-wide comparison, Germany is ranked on seventh place from the bottom. With respect to university graduates and management positions, the gap is even wider. One main reason for this gap is that women are still very rarely represented in certain professions, sectors and on the upper end of the job career ladder. As the wage gap is a key indicator of the persistent gender inequality in working life used in political and scientific debates, we chose this for the SIS. The ratio between women's and men's gross yearly earnings addresses nearly all problems women are still confronted with in their working lives: women's limited access to certain jobs, obstacles they face in their professional development, traditional gender roles and mental patterns which hamper the reconciling of family and working life, including obstacles to re-enter labour market after a career break due to child care. Each of these factors contributes to the pay gap, ultimately. An EU-wide comparison reveals that in Germany the gender pay gap in the sector electricity, gas, heat and cold supply belongs to those economic sectors with the highest gap [78].

In the highest salary group in the energy sector, women´s salary was in 2015 about 84 % of men's salary, with an annual salary difference of around 16,000 euros. It is linked to a number of legal, social and economic factors that go far beyond the single issue of equal pay for equal work. Nevertheless, the indicator provides a suitable benchmark to measure the achievement of equal opportunities within the energy sector. Official statistics distinguish between five performance groups representing a rough categorization of the employees' activities according to the qualification profile of workplaces. This categorization was narrowed down to the 'highest salary group' for a clearer visualization and focusing, and to ensure reliable data series from the Federal Statistical Office. This 'performance group 1' includes employees in a leading position with supervisory and discretionary authority such as employed managers, provided their earnings include, at least partially, non-performance-related payments. Employees in larger management areas who perform dispatching or management tasks are included as well as employees with activities that require comprehensive business or technical expertise. In general, the specialist knowledge is acquired through university studies.

The indicator selected is defined with respect to gross yearly income of fulltime employees in the energy supply sector including special payments, according to the German Federal Statistical Office category 'D – Energy supply', which includes electricity, gas, heat and cold supply sector [79]. Until 2030, the target is defined to eliminate this gap. The target for 2020 is determined by interpolating the average value of the last five years (2011–2015) and the zero target for 2030. The extrapolated trend calculated for 2011–2015 shows an increase of the gap by 24 % in 2020 compared to the average value over the years 2011 to 2015. This means that the indicator is assigned with a red traffic light and measures are required to reduce the gender pay gap in the highest salary group in the energy sector.

Grid expansion is essential for the security of supply, optimal use and efficient distribution of electricity from decentralized renewable energy sources. The Federal Network Agency plays a key role as a state authority in the field of electricity grid and is responsible in particular for the planning and approval of ultra-high voltage power lines. In all planning levels, the public, e.g. citizens, initiatives and interest groups, is involved in accordance with legal requirements by the relevant authorities. However, not all planning processes fulfill the qualitative standards for participation. Besides, many people believe that the existing degree of public participation in the planning processes for grid extension is neither appropriate nor sufficient. Against this background, the indicator 'Share of regulatory tools in the planning of power transmission grids which fulfill regulatory requirements' (see Factsheet No. 33) was defined. The indicator addresses the sustainability rule *'Participation in societal decision-making processes'* and refers to the involvement of the public into decision-making to improve the transparency and quality of decision-making processes and to generate more legitimation for decisions taken. Ultimately, this is expected to result in a higher acceptance or acceptability of energy infrastructure projects [80].

Energy technologies can be perceived as subjective impairment of recreational values, spiritual and sensual meanings or aesthetic contemplation potentials of nature [81]. The indicator 'Share of tourists who perceive energy power technologies as being disruptive in the vacation region' (see Factsheet No. 34) addresses the sensual perception of leisure travelers and tourists. It substantiates the sustainability rule *'Conservation of the cultural function of nature'*.

While there are ambitious government targets to increase the share of renewable energy, social acceptance is recognized to may be a constraining factor in achieving this target. This is particularly apparent in the case of wind energy and grid extension, which has become a subject of contested debates in Germany. The indicators 'Acceptance of renewable energies in the neighbourhood' (see Factsheet No. 35) and 'Acceptance of grid extension for achieving 100 % renewable energy supply' (see Factsheet No. 36) are proposed among the different facets of acceptance to address the rule *'Conservation of social resources'*. While there are ambitious government targets to increase the share of renewable energy in Germany, it is increasingly recognized that social acceptance of renewable energy technologies may be a constraining factor in achieving this target especially due to changes in land use and landscape that are associated with these technologies. This is particularly apparent in the case of wind energy, which has become a subject of contested debates mainly due to visual impacts of plants on characteristic landscapes. Apparently, contradictions exist between general public support for renewable energy innovation on the one hand, and obstruction or even resistance against the realization of specific projects in the neighbourhood, on the other hand. Against this background, the indicator 'Acceptance of renewable energies in the neighbourhood' was chosen for the SIS. Based on different surveys in various years, the acceptance of different elements of the energy system was analysed on behalf of the German Renewable Energies Agency [82]. Data are available for Germany for the years 2010 to 2016 [83–86]. As target for 2050, a total acceptance of renewable energy in the neighbourhood was assumed. Based on a linear interpolation between 100 % in 2050 and the average value for the past 5 years (2011–2015), the targets for 2020 (72 %) and 2030 (81 %) were determined. Compared to the average value for 2012 to 2016, the extrapolated trend calculated for the past 5 years (2012–2016) shows a decrease in the acceptance of renewable energy in the neighbourhood by 7.3 % in 2020. However, the target for 2020 requires an increase of 8.7 % compared to the average value of 2012 to 2016. Consequently, the indicator is rated with a red traffic light. The indicator 'Acceptance of grid extension for achieving 100 % renewable energy supply' is rated with a white traffic light since the time series of data required for an assessment is not yet available.

In addition to the acceptance of renewable energy in the neighbourhood in general, data are also available for the acceptance of specific renewable energy technologies, such as wind turbines, biomass plants, photovoltaic systems (solar parks), and for nuclear and coal-fired power plants. The percentages listed in Table 3.5 are based on regular surveys and represent the sum of positive answer options 'I like that' and 'I like that very much'. Looking at renewable energy technologies in more detail, biomass and wind energy plants experience the lowest level of social acceptance, whereas solar energy to produce electricity with photovoltaic panels in solar parks receive the highest level of acceptance (Table 3.5).

Table 3.5: Acceptance of renewable energy technologies in the neighbourhood (data from [83–86])

	Acceptance in the neighbourhood (%)						
	2010	**2011**	**2012**	**2013**	**2014**	**2015**	**2016**
Solar park	74	77	77	72	72	77	73
Biomass	4	36	36	39	39	39	38
Nuclear power plant	5	2	3	3	5	4	5
Coal-fired power plant	6	8	8	8	11	7	6
Wind turbines	56	60	61	59	61	59	52

3.4 Assessment of indicators to achieve substantial sustainability

Nine indicators operationalize the seven sustainability rules concerning the conditions to achieve the substantial rules (Table 3.6).

Table 3.6: Indicators, targets and assessment of the conditions to achieve the susbtantial rules

Sustainability rule	Sustainability indicator	Targets			Target references	Assess-ment
		2020	**2030**	**2050**		
Internalization of external social und ecological costs	37. Degree of internalization of energy-related external costs (%)	63	75	100	Target 2050 based on assumptions, linear extrapolation for 2030 and 2020.	○
Promotion of international co-operation	38. Share of development aid expenses on energy-related projects in relation to total GDP (%)	0.07	0.09	0.15	Target 2050 taken from [60] and linear extrapolation for 2030 and 2020.	● (green)
Society´s ability to respond	39. Share of households producing renewable electricity (%)	12	18	30	Targets based on assumptions.	○
	40. Share of households buying renewable electricity (%)	37	58	100	Target 2050 based on assumptions, linear extrapolation for 2030 and 2020.	○
Society´s ability of reflexivity	41. Share of installed smart meters mandatory for large electricity consumers (%)	22	48	100	Target 2050 based on assumptions, linear extrapolation for 2030 and 2020.	○
Society's capability of government	42. Volume of public-financed loans for energy-related investments (billion €)	23.7	27.3	31.4	Targets based on assumptions.	● (red)
Society's ability of self-organization	43. Number of energy cooperatives engaged in renewable energy plants	1,415	2,215	3,691	Targets based on assumptions.	● (yellow)
	44. Share of population living in regions with the objective to shift to 100 % renewable energy (%)	26	51	100	Target 2050 based on assumptions, linear extrapolation for 2030 and 2020.	○
Balance of power between societal actors	45. Share of the four biggest electricity companies on the market for the first-time sale of electricity (%)	≤ 60	≤ 60	≤ 60	Targets based on [87].	○

The rule *'Internalization of external social and ecological costs'* addresses the aspect that the degree of internalization for 2010, calculated with this approach, amounts to 48.9 % based on [88].

Activities related to the energy system often cause environmental impacts and according costs. External costs occur if producing or consuming energy services imposes costs upon third parties, such as air pollution related ecosystem or health impairment to individuals and according clean-up costs to the society. Therefore, internalisation of external costs aims at making such effects part of the decision-making process of energy providers and users, reducing occurring market failures and minimising negative impacts of the energy system on society's welfare. In order to estimate these costs, external effects of the energy system have been identified, assessed and monetised, as far as possible. Internalisation of external costs can be implemented by various policy measures, including market-based instruments (e.g. charges, taxes or tradable permits). Accordingly, fair and 'true' energy pricing is assumed to make it economically more attractive to both, using energy services with fewer negative environmental effects and healthcare costs, and reducing energy use in total, in order to bridge the gap between private and societal costs of energy production and use. This is why the authors have chosen the indicator 'Degree of internalization of energy-related external costs' for the SIS (see Factsheet No. 37). The degree of internalization of energy-related external costs is defined here as the coefficient between taxes on energy use (energy taxes, electricity taxes, motor vehicle taxes, air transport taxes, and road taxes) and environmental and healthcare costs due to electricity production and energy use for heating and transportation. Data are given for the years 2008 to 2010 and are calculated based on methodological guidance given in [64]. Therefore, taxes on air transport and on nuclear fuels, established since 2011, are so far not included in the methodology and the numbers presented. Data on energy taxes, electricity taxes and motor vehicle taxes are taken from [89] and [90], data on road taxes for trucks from [91], and data on environmental costs from [64].

According to [92], environmental costs resulting from the production of electricity in Germany include environmental and healthcare costs that result from direct emissions. Costs resulting from indirect emissions over the entire life cycle of energy production have also been taken into consideration. Since indirect emissions arise not only in Germany, EU cost rates have been considered as well. The costs of greenhouse gas emissions are determined as 80 € per t CO_2, including damage as well as abatement costs. Estimates of environmental and healthcare costs of nuclear energy differ widely within the literature available. Following the requirements of the Methodological Convention used here [92], the most expensive technology should be used for the calculations. In the case considered here, this is electricity production from lignite. Environmental costs of transportation include health effects, climate change effects, noise, and impact on nature and landscape, as well as effects caused by indirect emissions (construction, maintenance and disposal, fuel supply).

Total environmental costs, defined as described, amounted to 122.4 billion € in 2008, 115.2 billion € in 2009 and 120.6 billion € in 2010 [64]. In principle, data for other years can also be calculated by taking into consideration the mix of electricity production, heat energy consumption, as well as the relevant data for the transport sector for the different years. However, this is only reasonable if both the related environmental costs and the technologies (e.g., emission factors) do not change – an assumption that is not realistic. Thus, only calculations for other years are valid that takes into account such changes. Based on the methodology described, in 2010, the degree of internalization of external costs amounted to 48.9 % [64], [89], [90], [91]. An update beyond 2010 was not calculated because the results strongly depend on the development of emissions and the related healthcare costs. As target for 2050, a complete internalization of energy-related external costs was assumed. Based on a linear interpolation between 100 % in 2050 and the average value for the 3 years with data available (2008–2010), the targets for 2020 and 2030 were determined. A white traffic light was assigned to this indicator because no trendline and distance-to-target were calculated due to the lack of a sufficient data series.

The rule *'Promotion of international co-operation'* is addressed by the indicator 'Share of development aid expenses on energy-related projects relating to total gross domestic product' (see Factsheet No. 38). This indicator highlights to which extent Germany enhances international cooperation to facilitate access to clean energy research and technology, including renewable energy, energy efficiency and advanced and cleaner fossil-fuel technology, and promotes investment in energy infrastructure and clean energy technology.

The two indicators 'Share of households producing renewable electricity' (see Factsheet No. 39) and 'Share of households buying renewable electricity' (see Factsheet No. 40) are concerning the sustainability rule

'Society's ability to respond'. They highlight the ability of society to react to sustainability challenges caused by the energy system and to support energy transition processes. The energy transition provides new opportunities for citizens to respond and participate, e. g. by becoming producers of energy on their rooftops or by choosing a green electricity provider. By this, they can support the transition process and the use of local renewable resources. Photovoltaic (PV) systems installed by home owners are usually under 10 kWp. The systems within this power range make up less than 15 % of the total installed PV power in Germany, while large systems above 500 kWp make up about 30 % [93].

The indicator 'Share of installed smart meters mandatory for large electricity consumers' (see Factsheet No. 41) represents the rule *'Society's ability of reflexivity'*. Smart metering can contribute to a better control and the optimisation of the energy system. By providing and exchanging particular information about energy supply and demand, it can support an improved balancing of the mismatch between energy supply and demand over time which occurs in a system that is increasingly based on fluctuating wind and solar energy. A nationwide rollout of smart meters needs to be accompanied by a high standard regarding data protection.

The indicator 'Volume of public financed loans for energy-related investments' relates to the rule *'Society's capability of government'*. The indicator reveals the overall value of public loans granted for energy-related investments in private households and small and medium-sized companies (see Factsheet No. 42).

The rule concerning the 'Society's ability of self-organization' is substantiated by two indicators. The 'Number of energy cooperatives engaged in renewable energy plants' (see Factsheet No. 43) was selected to address civil societal activities, in particular by self-organized and membere-owned organization as well as by creating networks and participatory structures. Due to recent changes in the Renewable Energies Act existing energy cooperatives are increasingly withdrawing from new investment activities [94]. This indicates that energy cooperatives are not active in large-scale renewable energy production, due to both, the high upfront investment costs and the trend that the energy sector is being handed back gradually to big providers.

Various forms of energy cooperatives have been founded in Germany for more than a decade, allowing citizens to directly support the energy transition through own investments in and ownership of large-scale renewable energy plants that would be too expensive for single individuals alone, such as solar parks or wind turbines. To date, most energy cooperatives are formed at a local level, for example, by villagers investing in a nearby wind farm. The existence of a cooperative for local energy projects can contribute to a higher public acceptance of new systems to provide renewable energy. Furthermore, it is an indication of the self-organizational potential of a society to support the energy transition idea.

Information about energy cooperatives is taken from [94–96] and includes local and regional citizens' cooperatives. Here, only energy cooperatives under the umbrella of the Deutscher Genossenschafts- und Raiffeisenverband e.V. are taken into consideration. According to these studies, the accumulated number of energy cooperatives was 8 in 2006, 272 in 2010 and 812 in 2015, 129 new energy cooperatives were established in 2013. After changes in the German Renewable Energy Act (EEG), a tender procedure for renewable energy plants was established by which projects of energy cooperatives are disadvantaged systematically. For this reason, the number of new cooperatives decreased to 56 in 2014 and 40 in 2015. These figures may vary since some sources are based on the year of establishment, others on the year of registration. The contracts of these energy cooperatives include electricity production (87 % of all cooperatives in 2012 and 95 % in 2013), heat production (19 % in 2012, 16 % in 2013), grid operation (4 % in 2012 and 2013), and operation of district heating systems (20 % in 2012, 16 % in 2013). Since the results are based on a survey where multiple answers were possible, the added single percentages exceed the total of 100 % [94,95]. Civil power plants produced approximately 580 million kWh of renewable electricity in 2012 and 830 million kWh in 2013 [94,95].

No data series are available for the number of people belonging to these cooperatives. Only for 2011, it is confirmed that more than 80,000 citizens were engaged in energy cooperatives.

To preserve the ability for self-organization in the field of renewable energies, we derived the targets for 2020, 2030 and 2050 by assuming that the number of energy cooperatives should rise proportionately to the increase of the 'Share of renewable energy in gross final consumption of energy'.The extrapolated trend calculated based on the past 5 years (2011–2015) leads to an almost doubling until 2020 compared to the average value for 2011 to 2015. The target for 2020 (1,415 cooperatives) requires an increase of 112 % compared to the average values for the years 2011 to 2015 (666 cooperatives). This results in a deviation of 13 %, which was assigned with a yellow traffic light.

The indicator 'Share of population living in regions with the objective to shift to 100 % renewable energy' (see Factsheet No. 44) was selected because regions setting ambitious targets to shift their energy supply towards renewable energies support the energy transition by providing space to test innovative technologies and create new organizational forms of cooperation.

The rule *'Balance of power between societal actors'* requires the avoiding or reducing of high power concentration, asymmetric communication, and limited access to information and consultation. The indicator 'Share of the four biggest electricity companies on the market for the first-time sale of electricity' (see Factsheet No. 45) was selected to address this rule. As a result of the liberalization of the energy market in 1998, smaller companies and co-operatives producing renewable electricity entered the market. However, the German electricity sector is still characterized by a high degree of vertical and horizontal integration and dominated by the four large electricity provider RWE, E.ON, Vattenfall, and EnBW, which had a share of 76% on the market of first-time sale of electricity in 2015 [97].

3.5 Sustainability assessment of the German energy system

Table 3.7 gives an overview of the assessment results for all 45 indicators selected for the sustainability assessment of the German energy system. Only for 12 indicators, it can be assumed that the sustainability targets for 2020 can be achieved without additional policy measures (green traffic light). Four indicators are aligned with a yellow traffic light. However, severe political action is needed to reach the targets for 18 indicators assigned with a red traffic light. Another 11 indicators are assigned with a white traffic light due to the lack of available data series. The indicators and the assessment results are described in detail in the annex.

It can be noted, that indicators related to the maintenance of society's productive potential with regard to the use of renewable and non-renewable resources as well as environment pollution (No. 10 to 22), are all rated with a red traffic light, except the energy-related emissions of acid-forming gases (green), the modal split in the transport sector and the final energy productivity of the industry (yellow). The indicators assessing the sustainable development of human capital (No. 26 to 29), however, are evaluated with a green traffic light, except the indicator 'Numbers of start-ups' (No. 29).

Table 3.7: Indicator-based sustainability assessment of the German energy system

	Sustainability indicator	Assess-ment
Securing human existence	1. Energy-related emissions of particulate matter	🟢
	2. Energy-related emissions of cadmium	🟡
	3. Energy-related emissions of mercury	🟢
	4. Energy import dependency	🔴
	5. Monthly energy expenditures of households with a monthly net income less than 1,300 Euros	🟢
	6. SAIDI of electricity	🟢
	7. Relation of employees in the renewable energy sector to total employees	🔴
	8. Final energy consumption of private households per capita	🟢
	9. Relation of technician salary to manager salary at the big electricity suppliers	🔴
Maintaining society's productive potential	10. Share of renewable energy in gross final consumption of energy	🔴
	11. Area under cultivation of energy crops	🔴
	12. Unused renewable electricity due to management measures	🔴
	13. Use of primary energy	🔴
	14. Specific final energy consumption of households for heating (temperature-corrected)	🔴
	15. Final energy consumption in the transport sector	🔴
	16. Modal split in the transport sector	🟡
	17. Number of electric vehicles	🔴
	18. Final energy productivity of the German economy	🔴
	19. Final energy productivity of the industry	🟡
	20. Final energy productivity of trade, commerce and services	🔴
	21. Energy-related greenhouse gas emissions	🔴
	22. Energy-related emissions of acid-forming gases	🟢
	23. Energy-related hazardous solid wastes	⚪
	24. Amount of high-level radioactive waste which has not been transferred to a safe final disposal site	🔴
	25. Installed capacity of renewable energy power plants	🟢
	26. Number of university graduates in the field of energy sciences	🟢
	27. Federal expenditures for energy research	🟢
	28. Number of German patents in the field of renewable energy and energy efficiency	🟢
	29. Number of start-ups in the renewable energy and energy efficiency sector	🔴
	30. Added value creation from the renewable energy sector	🟢
	31. Added value creation from energy efficiency measures in households	⚪
Preserving society's options for development and action	32. Gender pay gap in the highest salary group in the energy sector	🔴
	33. Share of regulatory tools in the planning of power transmission grids that fulfil regulatory requirements	⚪
	34. Share of tourists who perceive energy power technologies as being disruptive in the vacation area	⚪
	35. Acceptance of renewable energies in the neighbourhood	🔴
	36. Acceptance of grid extension for achieving 100 % renewable energy supply	⚪
Conditions to achieve the substantial sustainability	37. Degree of internalization of energy-related external costs	⚪
	38. Share of development aid expenditure on energy-related projects in relation to total GDP	🟢
	39. Share of households producing renewable electricity	⚪
	40. Share of households buying renewable electricity	⚪
	41. Share of installed smart meters mandatory for large electricity consumers	⚪
	42. Volume of publicly funded loans for energy-related investments	🔴
	43. Number of energy cooperatives engaged in renewable energy plants	🟡
	44. Share of population living in regions with the objective to shift to 100 % renewable energy	⚪
	45. Share of the four biggest electricity companies on the market for the first-time sale of electricityConditions	⚪

As described before, it was not possible to define suitable indicators for all sustainability aspects affected by the energy transition. This was the case, for example, for the issue of preserving biodiversity. However, an impression of the status of the biodiversity could be given, using some indicators of SIS, whichmeasure driving forces considered as mainly responsible for the loss of biodiversity [98]. Some driving forces, such as the extent of land use, are listed in the SIS or can be translated into adequate indicators. This was done for the load of nutrients and pollutants that is referring to the indicators eutrophication and acidification and discharge of heavy metals (Table 3.8). Only one main driving force – the occurrence of invasive species – is not reflected in the SIS at all, because it is not related to the transition of the energy system.

Table 3.8: Indirect sustainability assessment of the impact of the energy system on biodiversity

Driving forces considered as mainly responsible for loss of biodiversity	**Assess-ment**
Land use, land use intensification and fragmentation (SI No. 11: Area under cultivation of area crops)	🔴
Eutrophication and acidification (SI No. 22: Energy-related emissions of acid-forming gases)	🟢
Global warming caused by greenhouse gas emissions (SI No. 21: Energy-related greenhouse gas emissions)	🔴
Pollution due to the discharge of heavy metals (SI. No. 2: Energy-related emissions of cadmium) (however the targets are related to human health not to the critical loads of biodiversity)	⚪
Pollution due to the discharge of heavy metals (SI No. 3: Energy-related emissions of mercury) (however the targets are related to human health not to the critical loads of biodiversity)	⚪
Overuse of natural resources (SI No. 10: Share of renewable energy in gross final consumption of energy)	🔴
Overuse of natural resources (SI No. 12: Unused renewable electricity due to management measures)	🔴

SI = Sustainability Indicator

Regarding the pollution of ecosystems due to the discharge of heavy metals, however, the critical loads concept should be used for the assessment rather than the emission values affecting the human health. For Germany, critical loads are available for Lead (Pb), Cadmium (Cd) and Mercury (Hg), taking into account both potential health effects and ecotoxic effects by measuring the maximum load of ecosystems. As a result of European mapping, critical loads exceedances in Germany are widespread for Pb and Hg, but hardly for Cd [99]. A review of these statements based on results of German deposition measurement networks in combination with dispersion models is not yet possible. For this reason, there are no spatially differentiated representations of critical loads for heavy metals by atmospheric immissions.

As shown in the overview of results in Table 3.8, in four out of seven indicators relevant for the preservation of biodiversity the 2020 targets will probably not be achieved. Thus, it can be noted that the energy system and its transition will continue to contribute to the loss of biodiversity. However, the targets for these indicators were not derived to address biodiversity aspects explicitly. Therefore, the statement is accordingly provisional and uncertain. Against this background, we recommend further research and empirical studies aiming at overcoming these limitations of measuring impacts of the energy system on biodiversity.

4 Discussion

The Sustainability Indicator Set (SIS) developed has similarities and overlaps to existing indicator sets such as the German monitoring report 'Energy of the future' [5], but provides also - in relation to other energy indicator systems - new indicators, mainly at the socio-technical interface of the energy system. The SIS is the result of normative decisions and selection processes, which included experts and stakeholders. Thus, the SIS has to be regarded as a sciene-basedpicture of the current state of knowledge and awareness that has to be adapted by society and politics and has to be improved over time.

Considering the number of 45 indicators included in the SIS may evoke the idea – most frequently expressed by politicians and decision-makers in economy and industry – to aggregate the single indicator assessment results to a kind of 'Sustainability index' for the energy system. The main argument behind this demand is to get quick information that can be communicated more easily. However, there is no scientifically proven approach to sum up such heterogeneous indicators as they are compromised in the SIS to generate a single sustainability score. Beyond that, an aggregated index would be of limited value for decision-makers, because recommendations for action have to address particular fields of action that cannot be identified based on an aggregated index, but need disaggregated information provided in terms of specific indicators and targets. The assessment with the SIS presents such information in a transparent format. In any case, users of the SIS may select indicators according to the specific context they are acting in.

Aside from the adequate selection and definition of indicators to address all sustainability goals, there is a discussion on the appropriate number of indicators for structuring or guiding political debates and decision-making processes. With respect to the number of indicators, there is a clear trade-off between an appropriate substantiation of sustainability aspects and the manageability of analyses, the communicability of results, and the applicability in decision-making processes. The final number of indicators (45) comprised in the SIS is regarded as a suitable compromise between manageability and depth of information. Sustainable development includes the idea of development, i.e. change. Thus, it is consequently important that the SIS has to be updated from time to time in order to keep its function as an assessment and alert system with regard to undesirable trends and changes. The developed SIS is designed well enough to fulfill these requirements and to be updated easily and cope with changes.

Beyond that, the assessment results show that over half of the assessed indicators (18 out of 34 assessed indicators) are rated with a red traffic light (see chapter 3.5). Comparing these results with the results of the German monitoring report reveals that our work comes up with a divergent set of indicators and divergent ratings for similar indicators. Based on these findings, the discussion focuses on the reasons for these differnces.

First, the applied process and results of defining new sustainability indicators at the socio-technical interface of the energy system are compared.

Then, the assessment results of selected indicators are discussed in comparision to existing evaluation results focusing on those indicators that are used in both, the SIS and the German monitoring report, but which show divergent assessment results.

The third section of the discussion is devoted to definition of targets for the years 2020, 2030 and 2050 and the impact of the target setting on the assessment results.

4.1 Comparison of sustainability indicator sets

The objective of the work was to provide knowledge for the scientific and political debate on indicators to assess the sustainability of the German energy system and its transition. This also includes considerations whether the sustainble development of the energy system can be evaluated comprehensively and sufficiently with the indicators proposed by the German monitoring report. Since we found that additional indicators are needed for a comprehensive assessment we proposed new sustainability indicators based on recent and current discussions and debates in science, policies and society. For the discussion of the SIS with its 45 indicators, we use only comparable indicators sets focusing on the same objective. For this purpose, the energy transition navigator developed by the Federation of German Industry (BDI) [28], the indicator sets of the Institute for Advanced Sustainability Studies [100] and the Fraunhofer Institute for Systems and Innovation Research [101] have been chosen as reference because all of them include initial approaches to define indicators at the socio-technical interface of the energy system. The BDI navigator considers the acceptance of the energy transition in general and of major projects in particular as well as rising energy prices, but doesn't address the distribution of benefits and burdens among different social groups or any participatory issues [28]. Besides, the proposed indicators (No. 34, 35 and 36) make a more detailed distinction of acceptance focusing on the hot spot of acceptance of the energy transition.

The indicators developed by the Fraunhofer Institute for Systems and Innovation Research include just one socio-technical indicator focusing on the energy-related expenses of households in relation to their monthly net income. This indicator is quite similar to indicator No. 5 of the SIS. In politics and science there seems to be consensus that such an indicator is eligible, if not necessary, because some households cannot afford to adequately heat their homes or pay their electricity bills [102]. Experts are assuming that up to 12 % of the households in Germany are vulnerable towards energy poverty [6]. However, there is no general agreement about the definition of an adequate indicator referring to this issue. This is partly due to the lack of information of how much energy is needed by low-income households to meet their demand. Evidence exists that they need relatively more energy because people usually spent more time at home due to 'mini-jobs' or unemployment, and use less energy efficient electric devices since they often cannot afford more efficient ones. Yet, the situation of energy poverty is not considered as dramatic due to the German welfare system which partly takes over the heating costs as well as the electricity expenses for benefit recipients [46]. Besides, there is a continuous amendment of social subsidies to compensate increasing energy prices and public consultation services are provided to help low-income households to improve the efficiency of their energy consumption and to support energy-related refurbishment.

The Institute for Advanced Sustainability Studies goes much further than just focusing on low-income households [100]. They propose indicators to monitor the distribution of benefits and costs among population groups at household level, as well as the collaborative aspects of the energy transition and the degree of commitment and participation. More specifically, they suggest recording the share of low-income households, which benefit from the feed-in profits of the Renewable Energy Act (EEG). This proposal is similar to our indicator No. 39. However, the focus here is directed at the degree of participation and not at economic benefits, which are considered not to be directly relevant for sustainable development. Another indicator proposed by Goldammer et al. [100] addressing energy poverty is the share of annually commissioned power cut-offs per 100 metering points. In Germany, more than 0.75 % of all households are going to have their power services cut-off each year because they could not pay their electricity bill [102]. In our work power cut-offs are not regarded as an adequate indicator for addressing energy poverty, because this parameter is influenced also by other factors than energy affordability. Besides, power cut-offs are prohibited in cases of households with kids or sick persons, while nevertheless such households might suffer from energy poverty.

The proposed indicators on gender pay gap (No. 32), the fulfilment of regulatory requirements (No. 33) as well as the indicators addressing the society's ability of response, reflexivity, governance and self-organization (No. 39 to 45) are not considered in any of the indicator sets mentioned above. This is remarkable, as there seem to be consensus in science and politics that active participation of citizens, for example by buying or producing renewable electricity or being part of energy cooperatives, is regarded as essential for a successful energy transition. Although the SIS was mainly compiled to improve existing indicator sets particularly at the socio-technical interface of the energy system, also new indicators for environmental and economic sustainability issues have been developed. An example is indicator No. 11. 'The area under cultivation of energy crops' is not part of the monitoring system or any of the indicator set mentioned, although increasing land use competition due to the energy transition has been a debated topic in science and politics for years.

For those indicators facing socio-technical aspects of the energy system (No. 32 to 45) only few data exist. Thus, in some cases, it is not possible yet to create data series of at least five years. Since the distance-to-target method applied requires such series, no assessment is possible for most of these indicators. Therefore, white traffic lights were assigned indicating the need to collect data that are more comparable over time. Since this is the case for 11 out of 45 indicators of the SIS, it is difficult to assess the social and socio-economic impacts of the energy system and its transition, being the field of investigation that is the most exciting from our point of view. Among the indicators related to the socio-technical interface, only one indicator is assigned with a green traffic light (No. 38), whereas three indicators (No. 32, 35 and 42) are assigned with a red traffic light. This indicates the need for action to close the gender pay gap in the energy sector, to increase public acceptance for renewable energies in the neighbourhood and also the volume of publicly funded loans for energy-related investments.

4.2 Comparison of assessment results

The quality and reliability of assessments based on the SIS depends on the appropriateness of the selected indicators, the availability of valid data series, targets determined, and the assessment method applied. These factors, their relationships and impacts on the assessment results will be discussed in the following by comparing our results mainly with those of the German monitoring report 'Energy of the Future', since this is the only official and the most elaborated and regularly revised approach to monitor the German Energiewende. Besides, it applies a similar approach for the selection of indicators for economic and ecological impacts and the assessment of the indicator performances. Other approaches, such as the Indicator report from the German Federal Office of Statistics or the Energiewende-Navigator, developed by the Federal Association of German Industry (see [28]), are not considered in this report, because they are not as comprehensive and regularly updated as the German monitoring report. Furthermore, they use another approach for the assessment resulting in another traffic light system that is not comparable with the approach described here.

The discussion is focusing on those indicators that are used in both the SIS and the German monitoring report [5], but show divergent assessment results. Such differences occur in the case of four indicators addressing key targets of the energy transition: the share of renewable energies in gross final energy consumption (No. 10), the primary energy use (No. 13), the final energy productivity of the German economy (No. 18), and the greenhouse gas emissions (No. 21). In our assessment, these indicators are all assigned with a red traffic light (see Table 4.1).

Although the monitoring report also applied the distance-to-target approach and the same data series (except for the greenhouse gas emissions where we included only the energy-related emissions), the two assessment results are different. This is related to the applied assessment rating system. In our rating system, a red traffic light is awarded, if the deviation between the projected value for 2020 and the target is 40 % or more. A yellow traffic light is given, if the deviation is between 10 % and 40 % and a green traffic light if the deviation is 10 % or less. Differing from this, the monitoring report applies an assessment scoring system ranging from 5 points for the fulfilment of a target with a deviation of up to 10 %, to 1 point for a deviation over 60 % (Table 4.1). Using this scoring method leads to the results that three of these four indicators (No. 13, 18 and 21) were awarded with 3 points, whereas the indicator No. 10 was awarded with 5 points. In fact, the monitoring report assessment results of these four indicators are more positive compared to the results presented here.This is, among others, because the evaluation approach of the monitoring report with a range of 5 points awards still 3 points when the distance to the target is between 20 and 40 percent. Löschel et al. who invoke that such great distances to the target can heavily be compensated for the year 2020 already criticized this [8]. They stated that missing the target of 2020 for the greenhouse gas emissions is likely and that this failure to meet the target is not reflected well enough by awarding 3 points. Furthermore, they highlighted the assessment of the final energy productivity (No. 18) as being excessively optimistic. Therefore, Löschel et al. also rate the final energy productivity and the greenhouse gas emissions with a red traffic light.

Table 4.1: Comparison of assessment results for selected indicators

Source of assessment	**SI No. 10:** Share of renewable energies in gross final consumption of energy	**SI No. 13:** Use of primary energy	**SI No. 18:** Final energy productivity of the German economy	**SI No. 21:** Energy-related greenhouse gas emissions
SIS assessment	🔴	🔴	🔴	🔴
Monitoring report 'Energy of the Future'[5]	5 points	3 points	3 points	3 points
Löschel et al. [8]	🟢	🟡	🔴	🔴

Another reason for the varying results is the methodology chosen to assess the deviation between projected values and the targets for the year 2020. As described before (see formula I in chapter 2.3), we compare the projected change in percentage with the change required in percentage for calculating the deviation in percentage that is evaluated using the traffic light colour code. In contrast, the monitoring report compares the absolute values of the projected value with the target. We chose the percentage deviation because it provides information on both, the deviation of the present and the projected value from the present and future target. Besides, absolute values could result in misleading conclusions. This applies particularly to cases where the distance between the current value and the target is large, because comparing absolute values would lead to an overestimation of the degree of target achievement. On the other hand, using percentage values as basis for the assessment can lead to an underestimation of the target achievement degree in cases where the distance between the current value and target is small.

Another methodological difference exists with respect to the reference value used for the calculation of the projected value for 2020. In the monitoring report, the projected value was derived by a linear projection starting from the year 2008, which is fixed for all indicators. In our assessment, however, we use the average value of the period of the past 5 years with available data. Although for many indicators data series up to the year 2015 or 2016 were available, this approach has the drawback that the indicators can have different reference periods. Despite this drawback we have chosen this approach in order to better capture and integrate recent changes in trend development, e.g. due to modifications of societal framework conditions, such as regulation approaches. To give an example: With just 40 new energy cooperatives being set up in 2015, the number of newly founded cooperatives decreased by another 25 % compared to the previous year with an already low level. Such recent shifts are possibly overlaid in the monitoring report, as has been already stated in [56]. Löschel et al. criticize the monitoring report being not able to suitably consider the more or less stagnation of greenhouse gas emissions since 2009 with its methodological approach. In contrast, we assigned a red traffic light to this indicator, as a result of regarding the probability to reach the target determined.

It has to be noted that the delimitation of the 5-year-period and the calculation of the reference value depends on the availability of data series. Consequently, the number of remaining years for political measures to achieve the 2020 target can differ. Considering a period closer to the target, e.g. from 2012 to 2016, would require stronger measures to achieve the target compared to an earlier time period, e.g. 2008 to 2012, because fewer years remain for interventions and measurable impacts. Thus, it may be reasonable to adjust the reference lines to assign the traffic light code over time. Moving closer to the target year 2020, the need for action is more urgent, and thus the traffic light should turn e. g. from a yellow light into a red light, accordingly a green traffic light could turn into a yellow one. Compared to the approach chosen, such a modification could better fulfil fairness considerations in the distance-to-target approach, but it would definitely make the assessment more complicated and require difficult decisions how to adjust the traffic light colour code in detail.

4.3 Comparison of targets and their impact on the assessment result

One important reason for the differences in the results between our assessment results and those in the monitoring report and the statement to the monitoring report [8] are the targets determined for the indicators. Löschel et al. assessed the indicator No. 13 ('Use of primary energy') with a yellow traffic light and the indicator No. 10 ('Share of renewable energy in gross final consumption of energy') with a green traffic light, meaning that it is likely that the targets for 2020 can be achieved with current policies and strategies. For indicator No. 10, we choose a more ambitious target for 2020 as Löschel et al. and the monitoring report. Instead of 18 % share of renewable energy, a share of 23 %, based on [40], was determined to ensure a better consistency with other assumptions also taken from [40]. Hence, we assigned the indicator No. 10 with a red traffic light, in contrast to the green traffic light in the monitoring report. This example shows the influence of target setting on the assessment results.

If available and possible, we have applied existing policy targets to be compatible and meaningful for political decision-makers and give applicable information. In view of the influence of the target definition on the assessment result, it can be criticised that targets should be defined according to scientific evidence rather than political feasibility. The debate on climate protection shows that this would probably lead to targets that are more ambitious and to a worse rating of the transformation strategies implemented. In our assessment, however, for many indicators this would not have changed the alignment of the already red traffic lights and the recommendation that action is required to reach the political targets.

For the new indicators that are not yet on the political agenda of the energy transition we have applied a scientific approach to derive appropriate targets for and beyond the year 2020. In the view of these findings, we consider it important for future research and according policy consultation to better consider strengths and weaknesses of sustainability assessments based on distance-to-target calculations, and also the impact of the selected reference values, targets defined and scoring systems applied on results and recommendations. One possibility to check and reveal the quality and robustness of assessment results could be to carry out sensitivity analyses to support decision-makers in becoming more aware how changes in reference values, distance-to-target calculations and targets can influence assessment results and policy recommendations.

In view of the long-term nature of the energy transition, the time horizon of the assessment has to be extended beyond 2020 and the progress in reaching the targets of the SIS has to be monitored periodically. The targets defined as well as the political action taken to reach the targets have to be reviewed to keep up with current and new developments that can occurr and might require a reinterpretation of certain targets. Additional efforts and measures could be necessary in a timely manner to avoid the otherwise foreseeable significant failure of targets set for the year 2030 and 2050. However, it should kept in mind that, depending on the extent to which the targets were missed by 2020, the achievement of the long-term targets for the year 2030 and 2050 will be further complicated.

5 Conclusions

The SIS developed within the Helmholtz Alliance ENERGY-TRANS is a comprehensive tool to assess progress towards a more sustainable energy system and is, thus, useful to support decision-making. It includes new indicators to assess the socio-technical interface of the system that are lacking in existing indicators sets such as the German monitoring report 'Energy of the Future'. The several new indicators that were proposed represent an innovative and major result of this work. They mainly address the interfaces between technology, society and economics and the collaborative design and development of the energy transition, which goes far beyond particular monetary aspects such as the costs for electricity supply. The SIS can help to reveal and eradicate the blind spots and weaknesses of existing indicator sets and to improve the assessment of issues at the socio-economic-technical interface of the energy system and its energy transition. Nevertheless, additional research and methodological work is required to further improve the SIS mainly with respect to sustainability issues neglected so far, such as the impact of the energy system and its transition on energy poverty.

The SIS is considered a relevant contribution to sustainability research and practice for the further development of the energy system and its transition. It can be used as a monitoring system by politics, administration, NGOs and society. As no other scientific approach provides a similar comprehensive tool for the sustainability assessment of energy systems, our work is a milestone that contributes both, to the academic discourse and the improvement of already existing indicator based assessments such as the German monitoring report. However, both the determination of indicators and targets as well as the assessment methodology should be seen as a continuous process in which scientists, decision-makers, stakeholders and citizens should be integrated. Since the distance-to-target methodology features some uncertainties and limitations that are associated with the method, it is crucial to check and display the quality and robustness of the assessment result by carrying out sensitivity analysis. As for over one quarter of the SIS no assessment was possible due to the lack of data series, research and monitoring it is highly recommended to start data collection and validation for these indicators in order to carry out a really comprehensive sustainability assessment. Another important issue for further research is the science-based definition of targets for the indicators. Since target setting is a process, which is subject to social value patterns and thus needs political agreement and legitimation, the process of target definition should include an inter- and transdiscioplinary approach. Nevertheless, the indicator system has to be updated from time to time to keep the function of the sustainability indicators as an alert system with regard to undesirable trends and reversals. The developed indicator system is designed well enough to be updated easily and cope with changes.

The SIS has the potential to provide information beyond the mere assessment of single indicators. For example, it is applicable to assess the impact on biodiversity in an indirect way and to identify trade-offs between sustainability issues. The assessment tool bears the potential for studying a wide range of questions concerning the future sustainability of the energy system. Besides, the SIS could be used to assess the sustainability of the energy system at different scales, at state level as well as in other European countries if data series are available. With respect to the methodological challenges outlined above, applying the SIS for monitoring and decision-making in different contexts and at different scales would be beneficial to gain experiences about the adaptability of the SIS assessment toll and to get valuable clues how to elaborate our approach.

References

[1] Federal Ministery of Economics and Technology (BMWi) and Federal Ministery for the Environment, Nature Conservation and Nuclear Safety (BMU)(Ed). 2010. Energy Concept for an Environmentally Sound, Reliable and Affordable Energy Supply. http://www.osce.org/eea/101047. Accessed: 29 August 2017

[2] Federal Ministry for Economic Affairs and Energy (BMWi)(Ed). 2012. A good piece of work – The Energy of the Future – First Energy Transition Progress Report. https://www.bmwi.de/Redaktion/EN/Publikationen/vierter-monitoring-bericht-energie-der-zukunft-kurzfassung.html. Accessed: 29 August 2017

[3] Federal Ministry for Economic Affairs and Energy (BMWi)(Ed). 2014. Second Monitoring Report 'Energy of the future.'. https://www.bmwi.de/Redaktion/DE/Publikationen/Energie/zweiter-monitoring-bericht-energie-der-zukunft.html. Accessed: 28 August 2017

[4] Federal Ministry for Economic Affairs and Energy (BMWi)(Ed). 2015. A good piece of work – The Energy of the Future – Fourth Energy Transition Monitoring Report. https://www.bmwi.de/Redaktion/EN/Publikationen/vierter-monitoring-bericht-energie-der-zukunft-kurzfassung.html. Accessed: 29 August 2017

[5] Federal Ministry for Economic Affairs and Energy (BMWi)(Ed). 2016. Fifth Monitoring Report 'The Energy of the Future'. https://www.bmwi.de/Redaktion/EN/Publikationen/monitoring-report-2016.html. Accessed: 29 August 2017

[6] Löschel, A.; Erdmann, G; Staiß, F.; Ziesing, H. 2014. Statement on the Second Monitoring Report of the Federal Government for 2012. https://www.bmwi.de/Redaktion/DE/Downloads/M-O/monitoring-bericht-energie-der-zukunft-stellungnahme-zusammenfassung-2014.pdf?__blob=publicationFile&v=5. Accessed: 29 August 2017

[7] Löschel, A.; Erdmann, G; Staiß, F.; Ziesing, H. 2015. Statement on the Fourth Monitoring Report of the Federal Government for 2014. https://www.bmwi.de/Redaktion/DE/Downloads/M-O/monitoring-bericht-energie-der-zukunft-stellungnahme-zusammenfassung-2014.pdf?__blob=publicationFile&v=5. Accessed: 29 August 2017

[8] Löschel, A.; Erdmann, G; Staiß, F.; Ziesing, H. 2016. Statement on the Fifth Monitoring Report of the Federal Government for 2015. https://www.bmwi.de/Redaktion/EN/Artikel/Energy/monitoring-implementation-of-the-energy-reforms.html. Accessed: 29 August 2017

[9] United Nations Conference on Sustainable Development (UNCSD)(Ed). 1996, 2001, 2007. The United Nations Indicators of Sustainable Development

[10] Renn, O. 2007. Leitbild Nachhaltigkeit. Eine normativ-funktionale Konzeption und ihre Umsetzung. VS Verlag für Sozialwissenschaften. Wiesbaden

[11] European Commission. 2007. Measuring progress towards a more sustainable Europe. 2007 monitoring report of the EU sustainable development strategy. EUR-OP. Luxembourg

[12] Tasaki, T.; Kameyama, Y.; Hashimoto, S.; Moriguchi, Y.; Harasawa, H. 2010. A survey of national sustainable development indicators. International Journal of Sustainable Development. 13. 4. 337

[13] Organisation for Economic Co-operation and Development (OECD)(Ed). 2000. Towards Sustainable Development. Indicators to measure progress. www.oecd.org/site/worldforum/33703694.pdf. Accessed: 29 August 2017

[14] United Nations (UN)(Ed). 2015. Sustainable Development Goals: 17 Goals to transform our world. http://www.un.org/sustainabledevelopment/sustainable-development-goals/. Accessed: 29 August 2017

[15] Kopfmüller, J. 2001. Nachhaltige Entwicklung integrativ betrachtet. Konstitutive Elemente, Regeln, Indikatoren. Ed. Sigma. Berlin

[16] Schultz, J.; Brand, F.; Kopfmüller, J.; Ott, K. 2008. Building a 'theory of sustainable development'. International Journal of Environment and Sustainable Development. 7. 4. 465

[17] Keiner, M. 2005. Re-emphasizing sustainable development. Environment, Development and Sustainability. 6. 4. 379–392

[18] World Commission on Environment and Development (WCED). Our Common Future. Oxford

[19] United Nations (UN). Report of the United Nations Conference on Environment and Development. A/CONf.151/26 (Vol. I-III). New York

[20] Minsch, J. 1998. Institutionelle Reformen für eine Politik der Nachhaltigkeit. Springer. Berlin u.a.

[21] Rösch, C.; Skarka, J.; Raab, K.; Stelzer, V. 2009. Energy production from grassland – Assessing the sustainability of different process chains under German conditions. Biomass and Bioenergy. 33. 4. 689–700

[22] Boulanger, P.-M. 2008. Sustainable development indicators. Surveys and Perspectives Integrating Environment and Society. 1. 1. 59–73

[23] United Nations (UN)(Ed). 2007. Indicators of Sustainable Development. Guidelines and Methodologies. www.un.org/esa/sustdev/natlinfo/indicators/guidelines.pdf. Accessed: 29 August 2017

[24] Bell, S.; Morse, S. 2010. Breaking through the Glass Ceiling. Local Environment. 6. 3. 291–309

[25] Barton, J.; Kopfmüller, J. (2012). Sustainable urban development in Santiago de Chile: Background - concept - challenges.In:Heinrichs, D.; Krellenberg, K.; Hansjürgen, B.; Martinez, F. Risk Habitat Megacity. Springer. Heidelberg, Dordrecht, London, New York

[26] Agentur für Erneuerbare Energien (AEE)(Ed). 2015. Bundesländer mit neuer Energie – Jahresreport Föderal Erneuerbar 2014/15. https://www.foederal-erneuerbar.de/bundeslaender-mit-neuer-energie-jahresreport-foederal-erneuerbar-2015. Accessed: 29 August 2017

[27] Agentur für Erneuerbare Energien (AEE); Deutsches Institut für Wirtschaftsforschung (DIW Berlin); Zentrum für Sonnenenergie und Wasserstoffforschung Baden-Württemberg (ZSW)(Eds). 2014. Vergleich der Bundesländer: Analyse der Erfolgsfaktoren für den Ausbau der Erneuerbaren Energien 2014 – Indikatoren und Ranking. https://www.diw.de/sixcms/detail.php?id=diw_01.c.489994.de. Accessed: 29 August 2017

[28] Bundesverband der Deutschen Industrie e.V.(Ed). 2014. BDI – Energiewende-Navigator 2014. Monitoring zur Umsetzung der Energiewende

[29] Brodmann, U.; Gubler, F.; Walter, F. 2001. Nachhaltigkeit: Kriterien und Indikatoren für den Energiebereich. www.bfe.admin.ch/php/modules/enet/streamfile.php?file=000000007094.pdf. Accessed 15 April 2017

[30] Compiling and Refining Environmental and Economic Account (CREEA). 2014. Compiling and refining environmental and economic accounts. Ergebnisberichte des EU-Projekts. Brüssel

[31] Statistisches Bundesamt (DESTATIS)(Ed). 2014. Nachhaltige Entwicklung in Deutschland - Indikatorenbericht 2014. https://www.destatis.de/DE/Publikationen/Thematisch/Umweltoekonomi-scheGesamtrechnungen/Umweltindikatoren/NationaleNachhaltigkeitsstrategie.html;jsessio-nid=2BD1FD4E1F6DFAF335B2684F48CB42F2.cae2. Accessed: 29 August 2017

[32] Statistisches Bundesamt - Destatis(Ed). 2016. Umweltökonomische Gesamtrechnungen. Nachhaltige Entwicklung in Deutschland Indikatoren zu Umwelt und Ökonomie. https://www.destatis.de/DE/Pub-likationen/Thematisch/UmweltoekonomischeGesamtrechnungen/Umweltindikatoren/Indikatoren-PDF_5850012.html. Accessed: 29 August 2017

[33] International Atomic Energy Agency (IAEA), UN Department for Economic and Social Affairs, International Energy Agency, Eurostat, European Environment Agency(Ed). 2005. Energy indicators for sustainable development. Guidelines and methodologies. http://www-pub.iaea.org/MTCD/publications/PDF/Pub1222_web.pdf. Accessed: 29 March 2017

[34] Goldammer, K.; Mans, U.; Rivera, M. 2013. Beiträge zur sozialen Bilanzierung der Energiewende. http://publications.iass-potsdam.de/pubman/faces/viewItemOverviewPage.jsp?itemId=escidoc:309117:5. Accessed: 29 August 2017

[35] Kearney, A. T. Energiewende-Index. WirtschaftsWoche

[36] Ministerium für Umwelt, Klima und Energiewirtschaft(Ed). Monitoring der Energiewende in Baden-Württemberg. Schwerpunkte Versorgungssicherheit und Effizienztrends Statusbericht 2015. https://um.baden-wuerttemberg.de/fileadmin/redaktion/m-um/intern/Dateien/Dokumente/2_Presse_und_Service/Publikationen/Energie/Monitoring_Energiewende_2015.pdf. Accessed: 30 August 2017

[37] Flues, F.; Löschel, A.; Pothen; F.; Wölfing, N. 2012. Indikatoren für die energiepolitische Zielerreichung. ftp://ftp.zew.de/pub/zew-docs/gutachten/ZEW_Indikatorenbericht_2012.pdf. Accessed: 30 August 2017

[38] Burgherr, P.; Hirschberg, S. 2014. Comparative risk assessment of severe accidents in the energy sector. Energy Policy. 74. S1. 45–56

[39] Burgherr, P.; Hirschberg, S.; Spada, M. (2013). Comparative Assessment of Accident Risks in the Energy Sector.In:Kovacevic, RM.; Pflug, G.; Vespucci, M. Handbook of risk management in energy production and trading. Springer US. Boston, MA, s.l.

[40] Deutsches Zentrum für Luft- und Raumfahrt; Ingenieurbüro für neue Energien (DLR); Fraunhofer Institut für Windenergie und Energiesystemtechnik (IWES); Ingenieurbüro für neue Energien (IFNE)(Eds). Langfristszenarien und Strategien für den Ausbau der erneuerbaren Energien in Deutschland bei Berücksichtigung der Entwicklung in Europa und global. Schlussbericht BMU – FKZ 03MAP146. http://www.dlr.de/dlr/Portaldata/1/Resources/bilder/portal/portal_2012_1/leitstudie2011_bf.pdf. Accessed: 14 June 2016

[41] Pye, S.; Dobbins, A. 2015. Energy poverty and vulnerable consumers in the energy sector across the EU: analysis of policies and measures

[42] Council of European Energy Regulators (CEER). 2015. CEER Benchmarking Report 5.2 on the Continuity of Electricity Supply. Data update, 12 February 2015

[43] Statistisches Bundesamt (DESTATIS)(Ed). Number of employees 2006 to 2015. https://www.destatis.de/DE/ZahlenFakten/Indikatoren/Konjunkturindikatoren/Arbeitsmarkt/karb811.html. Accessed: 20 March 2017

[44] Bundesministerium für Wirtschaft und Technologie, Bundesministerium für Umwelt, Naturschutz und Reaktorsicherheit(Ed). 2010. Energiekonzept für eine umweltschonende, zuverlässige und bezahlbare Energieversorgung. https://www.bmwi.de/Redaktion/DE/Downloads/E/energiekonzept-2010.html. Accessed: 19 October 2017

[45] Zumach, A. Volksabstimmung in der Schweiz. Topverdiener werden nervös. taz. 9. http://www.taz.de/!5066916/. 23. May 2013

[46] Schweizer-Ries, P. 2009. Energy poverty: impact and public recognition in the United Kingdom, France, Germany, Italy and Poland. Project report within the scope of FinSH 'Financial and Support Instruments for Fuel Poverty in Social Housing

[47] Heindl, P. 2015. Measuring Fuel Poverty: General Considerations and Application to German Household Data. Discussion Paper No. 13-046. ftp://ftp.zew.de/pub/zew-docs/dp/dp13046.pdf. Accessed: 20 March 2017

[48] Tews, K. 2013. Energiearmut definieren, identifizieren und bekämpfen. FFU-Report 04-2013

[49] Statistisches Bundesamt (DESTATIS). 2015. Wirtschaftsrechnungen: Einkommens- und Verbrauchsstichprobe – Einnahmen und Ausgaben privater Haushalte für den Privaten Konsum – 2013. Fachserie 15 Heft 5. https://www.destatis.de/DE/Publikationen/Thematisch/EinkommenKonsumLebensbedingungen/Konsumausgaben/EVS_AufwendungprivaterHaushalte2152605139004.pdf?__blob=publicationFile. Accessed: 20 March 2017

[50] Statistisches Bundesamt (DESTATIS)(Ed). Laufende Wirtschaftsrechnungen – Einkommen, Einnahmen und Ausgaben privater Haushalte 2015. Fachserie 15 Reihe 1. https://www.destatis.de/DE/Publikationen/Thematisch/EinkommenKonsumLebensbedingungen/EinkommenVerbrauch/Einnahmen-AusgabenprivaterHaushalte2150100157004.pdf?__blob=publicationFile. Accessed: 20 March 2017

[51] Statistisches Bundesamt (DESTATIS)(Ed). Laufende Wirtschaftsrechnungen – Einnahmen und Ausgaben privater Haushalte 2012. Fachserie 15 Reihe 1. https://www.destatis.de/DE/Publikationen/Thematisch/EinkommenKonsumLebensbedingungen/EinkommenVerbrauch/EinnahmenAusgabenprivaterHaushalte2150100127004.pdf?__blob=publicationFile. Accessed: 20 March 2017

[52] Statistisches Bundesamt (DESTATIS)(Ed). Laufende Wirtschaftsrechnungen – Einkommen, Einnahmen und Ausgaben privater Haushalte 2014. Fachserie 15 Reihe 1. https://www.destatis.de/DE/Publikationen/Thematisch/EinkommenKonsumLebensbedingungen/EinkommenVerbrauch/EinnahmenAusgabenprivaterHaushalte2150100147004.pdf?__blob=publicationFile. Accessed: 20 March 2017

[53] Boardman, B. 1991. Fuel poverty. From cold homes to affordable warmth. Belhaven Press. London

[54] Umweltbundesamt(Ed). 2017. Private Haushalte und Konsum. https://www.umweltbundesamt.de/daten/private-haushalte-konsum/energieverbrauch-privater-haushalte. Accessed: 20 March 2017

[55] Lehr, U.; Edler, D.; O'Sullivan, M.; Peter, F.; Bickel, P. 2015. Beschäftigung durch erneuerbare Energien in Deutschland: Ausbau und Betrieb, heute und morgen. Studie im Auftrag des Bundesministeriums für Wirtschaft und Energie. https://www.zsw-bw.de/uploads/media/Studie_Beschaeftigung_durch_EE_2015.pdf. Accessed: 20 March 2017

[56] O'Sullivan, M.; Lehr, U.; Edler, D. Bruttobeschäftigung durch erneuerbare Energien in Deutschland und verringerte fossile Brennstoffimporte durch erneuerbare Energien und Energieeffizienz. Stand September 2016. https://www.bmwi.de/Redaktion/DE/Downloads/S-T/bruttobeschaeftigung-erneuerbare-energien-monitioringbericht-2015.pdf?__blob=publicationFile&v=11. Accessed: 12 April 2017

[57] indexmundi(Ed). Elektrizität Verbrauch pro Kopf - Welt. https://www.indexmundi.com/map/?v=81000&l=de. Accessed: 07 March 2017

[58] statista Datenbank(Ed). Anbaufläche von Energiepflanzen in Deutschland nach Art in den Jahren 2007 bis 2015. https://de.statista.com/statistik/daten/studie/153072/umfrage/anbauflaeche-von-energiepflanzen-in-deutschland-nach-sorten-seit-2007/. Accessed 30 August 2017

[59] Fern, WWF, Birdlife, Greenpeace, OXFAM, Transport&Environment, EE, actionaid(Ed). 2015. Pitfalls and potentials - The role of bioenergy in the EU climate and energy policy post 2020. NGO recommendations. http://www.wwf.eu/what_we_do/climate/climate_energy_publications.cfm?245290/Pitfalls-and-potentials---The-role-of-bioenergy-in-the-EU-climate-and-energy-policy-post-2020-NGO-recommendations. Accessed 27 April 2017

[60] Wissenschaftlicher Beirat der Bundesregierung Globale Umweltveränderungen(Ed). 2003. Welt im Wandel – Energiewende zur Nachhaltigkeit. http://www.wbgu.de/fileadmin/user_upload/wbgu.de/templates/dateien/veroeffentlichungen/hauptgutachten/jg2003/wbgu_jg2003.pdf. Accessed: 19 October 2017

[61] Lexas. Flächendaten aller Staaten der Erde(Ed). http://www.laenderdaten.de/geographie/flaeche_staaten.aspx. Accessed: 30 November 2016

[62] Umweltbundesamt(Ed). 2017. Strategien zur Emissionsminderung von Luftschadstoffen. https://www.umweltbundesamt.de/daten/luftbelastung/massnahmen-zur-emissionsminderung-von. Accessed: 11 November 2017

[63] Umweltbundesamt(Ed). 2017. Nationale Trendtabellen für die deutsche Berichterstattung atmosphärischer Emissionen 1990–2015. https://www.umweltbundesamt.de/themen/klima-energie/treibhausgas-emissionen. Accessed 30 August 2017

[64] Umweltbundesamt(Ed). 2015. Daten zur Umwelt 2015. https://www.umweltbundesamt.de/sites/default/files/medien/376/publikationen/daten_zur_umwelt_2015.pdf. Accessed: 20 March 2017

[65] Hirschl, B.; Heinbach, K.; Prahl, A.; Salecki, S.; Schröder, A.; Aretz, A.; Weiß, J. Dezember 2015. Wertschöpfung durch erneuerbare Energien. Ermittlung der Effekte auf Länder- und Bundesebene. Institut für ökologische Wirtschaftsforschung Berlin. Berlin

[66] Prahl, A. 2014. Renewable energies' impact on value added and employment in Germany – Model results for 2012. Conference presentation.

[67] Statistisches Bundesamt (DESTATIS)(Ed). GENESIS-Online Datenbank. https://www-genesis.destatis.de/genesis/online/data;jsessionid=4B5CA2EE9B7FD908DA4199F1BADA9DF6.tomcat_GO_2_2?operation=abruftabelleAbrufen&selectionname=21321-0003&levelindex=1&levelid=1430750648551&index=3. Accessed: 30 August 2017

[68] Bundesministerium für Wirtschaft und Energie(Ed). 2017. Bundesbericht Energieforschung 2017. Forschungsförderung für die Energiewende. [1] Bundesministerium für Wirtschaft und Energie: Bundesbericht Energieforschung 2016. Forschungsförderung für die Energiewende. 2017, p 42, https://www.ptj.de/lw_resource/datapool/_items/item_7306/bundesbericht-energieforschung-2016.pdf. Accessed: 30 August 2017

[69] Nagaoka, S.; Motohashi, K.; Goto, A. 2010. Patent Statistics as an Innovation Indicator. Handbook of the Economics of Innovation. 1083–1127

[70] Harhoff, D.; Scherer, F. 2000. Technology policy for a world of skew-distributed outcomes. Research Policy. 29. 4-5. 559–566

[71] Wuebker, R.; Wüstenhagen, R. 2011. The handbook of research on energy entrepreneurship. Edward Elgar. Cheltenham, UK, Northampton, MA

[72] Farla, J.; Markard, J., Raven, R.; Coenen, L. 2012. Sustainability transitions in the making: A closer look at actors, strategies and resources. Technological Forecasting and Social Change. 991–998

[73] Santos, F.; Eisenhardt, K. 2009. Constructing Markets and Shaping Boundaries: Entrepreneurial Power in Nascent Fields. The Academy of Management Journal. 643–671

[74] Borderstep Institut für Innovation und Nachhaltigkeit. 2015. Green Economy Gründungsmonitor 2014 – Grüne Wirtschaft als Gründungs- und Beschäftigungsmotor in Deutschland. http://www.borderstep.de/wp-content/uploads/2015/05/Green_Economy_Gruendungsmonitor_20141.pdf. Accessed: 20 March 2017

[75] Zentrum für Europäische Wirtschaftsforschung (ZEW)(Ed). 2014. Potenziale und Hemmnisse von Unternehmensgründungen im Vollzug der Energiewende. Studie im Auftrag des Bundesministeriums für Wirtschaft und Energie. ftp.zew.de/pub/zew-docs/gutachten/Projektbericht_Energiewende2014.pdf. Accessed: 20 March 2017

[76] Ausgaben für Forschung und Entwicklung in Prozent des BIP in ausgewählten Ländern im Jahr 2013. https://de.statista.com/statistik/daten/studie/158150/umfrage/ausgaben-fuer-forschung-und-entwicklung-2008/. Accessed: 30 March 2016

[77] Ministry of Family Affairs, Senior Citizens, Women and Youth (BMFSFJ)(Ed). Pay Inequality between Women and Men in Germany. https://www.bmfsfj.de/blob/94442/efbd528467e361882848c23486fcc8d8/pay-inequality-data.pdf. Accessed: 20 March 2017

[78] Eurostat(Ed). 2017. Gender pay gap statistics. http://ec.europa.eu/eurostat/statistics-explained/index.php/Gender_pay_gap_statistics. Accessed: 20 March 2017

[79] Statistisches Bundesamt (DESTATIS). Verdienste und Arbeitskosten. 2. 1. Arbeitnehmerverdienste 2007 – 2016. Fachserie. 16. https://www.destatis.de/GPStatistik/receive/DESerie_serie_00000297?list=all.
Accessed: 20 March 2017

[80] Schweizer, P-J.; Renn, O. Partizipation in Technikkontroversen: Panakeia für die Energiewende? Technikfolgenabschätzung – Theorie und Praxis. 22. 2

[81] Gallego, CD.; Mack, A. 2. Sustainability assessment of energy technologies via social indicators: Results of a survey among European energy experts. Energy Policy. 38. 1030–1039

[82] Agentur für Erneuerbare Energien (AEE)(Ed). Bundesländer-Übersicht zu Erneuerbaren Energien. https://www.foederal-erneuerbar.de/uebersicht/bundeslaender/BW%7cBY%7c%20B%7cBB%7cHB%7cHH%7cHE%7cMV%7cNI%7cNRW%7cRLP%7cSL%7cSN%7cST%7cSH%7cTH%7cD/kategorie/akzeptanz/. Accessed: 15 January 2018

[83] Agentur für Erneuerbare Energien (AEE)(Ed). Umfrage 2013: Bürger befürworten Energiewende und sind bereit, die Kosten dafür zu tragen. http://www.unendlich-viel-energie.de/themen/akzeptanz2/akzeptanz-umfrage/umfrage-2013-buerger-befuerworten-energiewende-und-sind-bereit-die-kosten-dafuer-zu-tragen. Accessed: 20 March 2017

[84] Agentur für Erneuerbare Energien (AEE)(Ed). Umfrage Akzeptanz Erneuerbarer Energien 2014. http://www.unendlich-viel-energie.de/mediathek/grafiken/akzeptanzumfrage-erneuerbare-energie-2014. Accessed: 20 March 2017

[85] Agentur für Erneuerbare Energien (AEE)(Ed). Umfrage zur Akzeptanz Erneuerbarer Energien 2015. http://www.unendlich-viel-energie.de/mediathek/grafiken/umfrage-akzeptanz-erneuerbare-energien-2015. Accessed: 20 March 2017

[86] Agentur für Erneuerbare Energien (AEE)(Ed). Umfrage 2016: Bürger befürworten Energiewende und sind bereit, die Kosten dafür zu tragen. https://www.unendlich-viel-energie.de/mediathek/grafiken/akzeptanz-umfrage-2016. Accessed: 20 March 2017

[87] Lang, C. 2008. Marktmacht und Marktmachtmessung im deutschen Großhandelsmarkt für Strom. Gabler Verlag

[88] Umweltbundesamt(Ed). 2017. Emissionen von Luftschadstoffen – Emissionsentwicklung 1990–2015 für klassische Luftschadstoffe. https://www.umweltbundesamt.de/themen/luft/emissionen-von-luftschadstoffen.Accessed 31 August 2017

[89] Bundesfinanzministerium(Ed). Kassenmäßige Steuereinnahmen nach Steuerarten in den Kalenderjahren 2006 – 2009. http://www.bundesfinanzministerium.de/Content/DE/Standardartikel/Themen/Steuern/Steuerschaetzungen_und_Steuereinnahmen/2012-05-29-steuereinnahmen-nach-steuerarten-2006-2009.pdf?__blob=publicationFile&v=3.
Accessed: 22 March 2017

[90] Bundesfinanzministerium(Ed). Kassenmäßige Steuereinnahmen nach Steuerarten in den Kalenderjahren 2010 – 2015. http://www.bundesfinanzministerium.de/Content/DE/Standardartikel/Themen/Steuern/Steuerschaetzungen_und_Steuereinnahmen/2016-05-24-steuereinnahmen-nach-steuerarten-2010-2015.pdf?__blob=publicationFile&v=4. Accessed: 22 March 2017

[91] Statista. Mauteinnahmen in Deutschland von 2005 bis 2015* (in Milliarden Euro). http://de.statista.com/statistik/daten/studie/75600/umfrage/mauteinnahmen-in-deutschland-seit-2005/. Accessed: 20 March 2017

[92] Umweltbundesamt(Ed). 2013. Ökonomische Bewertung von Umweltschäden – Methodenkonvention 2.0 zur Schätzung von Umweltkosten. https://www.umweltbundesamt.de/publikationen/oekonomische-bewertung-von-umweltschaeden-0. Accessed: 20 March 2017

[93] Fraunhofer ISE(Ed). 2016. Recent Facts about Photovoltaics in Germany. https://de.scribd.com/document/332431802/2016-Apr-22-Recent-Facts-About-PV-in-Germany. Accessed: 31 August 2017

[94] Deutscher Genossenschafts- und Raiffeisenverband e.V. (DRGV)(Ed). 2014. Energiegenossenschaften. Ergebnisse der Umfrage des DGRV und seiner Mitgliedsverbände. http://www.begeb.de/mediapool/139/1396625/data/Auswertung20Studie20BroschC3BCre202014.pdf. Accessed: 20 March 2017

[95] Deutscher Genossenschafts- und Raiffeisenverband e.V. (DRGV)(Ed). Energiegenossenschaften. Ergebnisse der Umfrage des DGRV und seiner Mitgliedsverbände. Frühjahr 2013. https://www.dgrv.de/webde.nsf/7d5e59ec98e72442c1256e5200432395/dd9db514b5bce595c1257bb200263bbb/$FILE/Umfrageergebnisse%20Energiegenossenschaften.pdf. Accessed: 10 April 2017

[96] Deutscher Genossenschafts- und Raiffeisenverband e.V. (DRGV)(Ed). 2015. Energiegenossenschaften. Ergebnisse der DGRV-Umfrage (zum 31.12.2015). https://www.genossenschaften.de/sites/default/files/Auswertung%20Jahresumfrage_0.pdf. Accessed: 10 April 2017

[97] Bundesnetzagentur, Bundeskartellamt(Ed). 2017. Monitoring report 2016. https://www.bundesnetzagentur.de/SharedDocs/Downloads/DE/Sachgebiete/Energie/Unternehmen_Institutionen/DatenaustauschUndMonitoring/Monitoring/Monitoringbericht2016.pdf?__blob=publicationFile&v=2. Accessed: 31 August 2017

[98] Umweltbundesamt(Ed). 2014. Biodiversität. https://www.umweltbundesamt.de/das-uba/was-wir-tun/forschen/umwelt-beobachten/biodiversitaet#textpart-1. Accessed: 29 March 2017

[99] Umweltbundesamt(Ed). Critical Loads für Schwermetalle. https://www.umweltbundesamt.de/themen/luft/wirkungen-von-luftschadstoffen/wirkungen-auf-oekosysteme/critical-loads-fuer-schwermetalle. Accessed: 15 January 2018

[100] Goldammer, K., Mans, U., Rivera M. 2013. Beiträge zur sozialen Bilanzierung der Energiewende. http://publications.iass-potsdam.de/pubman/item/escidoc:309117:7/component/escidoc:466896/IASS_Report_309117.pdf. Accessed: 15 January 2018

[101] Schlomann, B.; Reuter, M.; Eichhammer, W.; Ziesing H-J. 2016. Methoden- und Indikatorenentwicklung für Kenndaten zum Klimaschutz im Energiebereich. https://www.umweltbundesamt.de/publikationen/methoden-indikatorenentwicklung-fuer-kenndaten. Accessed: 30 August 2017

[102] Bundesnetzagentur(Ed). 2015. Monitoring report. https://www.bundesnetzagentur.de/DE/Sachgebiete/ElektrizitaetundGas/Unternehmen_Institutionen/DatenaustauschundMonitoring/Monitoring/Monitoringberichte/Monitoring_Berichte_node.html. Accessed 30 August 2017

Appendix: Indicator Factsheets

The factsheets are split into nine different categories of information in the following order:

(1) Description of the sustainability rule, which is addressed by the indicator and justification why the indicator was selected to represent this rule

(2) Definition of the indicator and comments on the definition of the indicator

(3) Reference unit

(4) Data series of the previous 10 years (if available), calculated trend until the year 2020 based on the previous 5 years with data (if available)

(5) Definition of targets for 2020, 2030 and 2050

(6) Assessment of the trend for the year 2020 and allocation of a traffic light symbol

(7) Comments on data used

(8) Comments on targets defined

(9) Literature used

1. Energy-related Emissions of Particulate Matter

Justification referring to Sustainability Concept	This indicator addresses the sustainability rule *'Protection of human health'*: Protecting every individual's life and health is an existential issue of sustainability based on the precautionary principle. The aim is to avoid health problems and risks actually or potentially caused by the energy system, e.g., noise, emissions of air pollutants and particulate matter, radioactive radiation, etc.
Description	Particulate matter is a complex mixture of solid and liquid particles and is divided into different groups according to their size. Differentiation is made between PM10 (PM, particulate matter) with a diameter of less than 10 micrometers (µm), PM2.5 with a maximum diameter of 2.5 µm, and ultrafine particles with a diameter less than 0.1 µm.
Unit	kt/a
Data	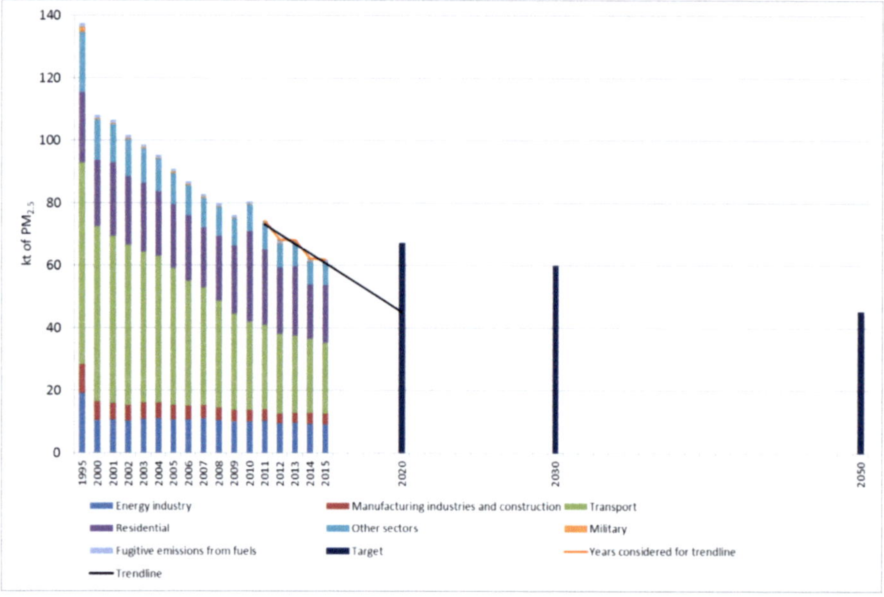 Figure 1: Energy-related emissions of particulate matter <2.5 µm (own diagram based on [1]) only direct emissions, without upstream and downstream values from [1])
Targets	2020: 67.4 kt 2030: 60.1 kt 2050: 45.6 kt
Assessment	 Value for the reference year (2005): 91.1 kt [1] Latest available value (2015): 61.8 kt [1] The target is a reduction of 0.6 % by 2020 compared to the refrence value (the average value over the previous 5 years 2011–2015). The trend calculated based on the previous 5 years would even result in a reduction of 33 % by 2020 ('projected value' for 2020). Therefore, so far this indicator is on track to reach the target for the year 2020, which results in a green light.

Comments Data	The term particulate matter subsumes directly emitted and secondarily formed particulate matter. Primary particulate matter is emitted directly at the source by emissions from motor vehicles, power and district heating plants, domestic heating systems, metal and steel production, or during the handling of bulk materials. But it can also be of natural origin (e.g., as a result of soil erosion). In urban areas, road transport is the dominant source of dust. Here, particulate matter are emitted into the air not only from engines – primarily from diesel engines – but also by brake and tire wear and the resuspension of road dust. Another important source is agriculture [2,3]. Among the particulate matter, PM2.5 is particularly harmful to health because it enters into the bronchi and alveoli or even the bloodstream of the human body. The health effects of particulate matter depend on particle size and penetration depth. They range from mucosal irritation and local inflammation in the trachea and the bronchi or the alveoli to increased plaque formation in the blood vessels, an increased risk of thrombosis, or changes in the regulatory function of the autonomic nervous system (heart rate variability) [2,3]. The corresponding data for PM2.5 emissions were taken from the data sheets of the Federal Environment Agency (UBA – Umweltbundesamt) [1]. These data include only direct emissions. Indirect emissions as they arise, for example, in the production of plants and emissions abroad from imported energy are not taken into account.
Comments Targets	For the protection of human health, a European target value of 25 µg/m3 for annual mean PM2.5 concentrations was set in 2008 and entered into force on 1 January 2010. This value is set as a limit value to be met by 1 January 2015, and as of 1 January 2020, annual mean PM2.5 concentrations must not exceed the value of 20 µg/m3. The Federal Environment Agency published the 2015 annual mean PM2.5 concentrations for 175 stations in or near urban areas. The mean value of all stations was 13.1 µg/m3. At two stations, the measured values were close to the limit value (19 µg/m3); at 39 stations, they were >= 15µg/m3 [3]. Since a large number of concentration values are close to the limit value to be met by 2020, a reduction in emissions is deemed necessary.The amended protocol of Gothenburg, which is based on the Convention on Long-range Transboundary Air Pollution in Geneva, states that Germany has to reduce its PM2.5 emissions by 26 % by the year 2020 compared to 2005 [2]. Using the same reduction target for energy-related PM2.5 emissions, this means a reduction from 91.1 kt in 2005 to 67.4 kt in 2020. Because of the high health risks of particulate matter, the value should be further reduced after 2020. A reduction in the value for the reference year 2005 of 50 % by 2050 is considered appropriate for reasons of public health. This means that in 2050 a value of 45.6 kt and in 2030 of 60.1 kt should not be exceeded. The value for the year 2030 is the result of a linear interpolation between the values for 2020 and 2050.
Literature	[1] UBA – Umweltbundesamt: Emissionen von Luftschadstoffen. Emissionsentwicklung 1990–2015 für klassische Luftschadstoffe. https://www.umweltbundesamt.de/themen/luft/emissionen-von-luftschadstoffen, access 8 March 2017 [2] UBA – Umweltbundesamt: Daten zur Umwelt 2015. http://www.umweltbundesamt.de/publikationen/daten-zur-umwelt-2015, access 6 August 2015 [3] UBA – Umweltbundesamt: Feinstaub. 11.3.2016. http://www.umweltbundesamt.de/themen/luft/luftschadstoffe/feinstaub, access 5 April 2016

2. Energy-related Emissions of Cadmium

Justification referring to Sustainability Concept	This indicator addresses the sustainability rule *'Protection of human health'*: Protecting every individual's life and health is an existential issue of sustainability based on the precautionary principle. The aim is to avoid health problems and risks actually or potentially caused by the energy system, e.g., noise, emissions of air pollutants and particulate matter, radioactive radiation, etc.
Description	Cadmium is toxic if it enters the human body in too high quantities. It is mainly ingested with the food.
Unit	t/a
Data	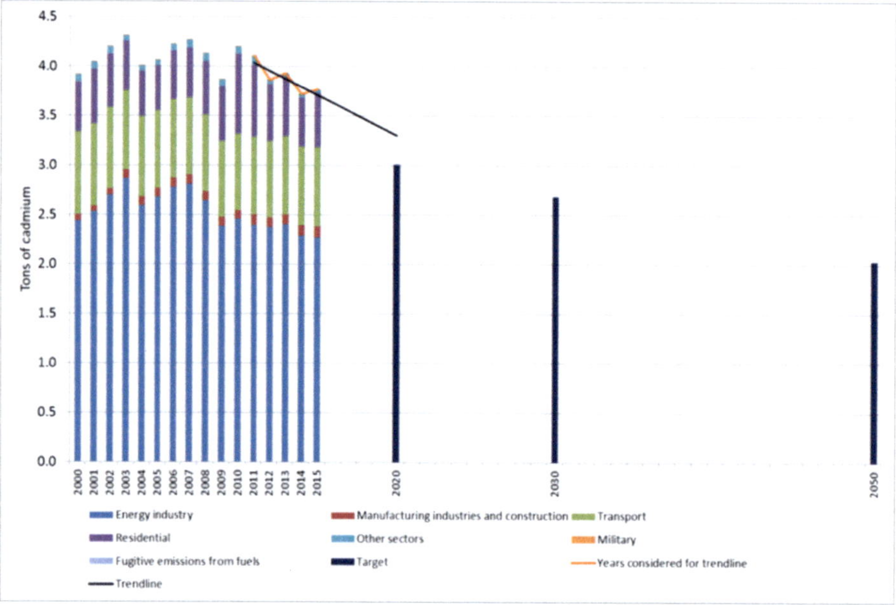 Figure 1: Energy-related emissions of cadmium (own diagram based on [1]) In 2015, energy-related emissions of cadmium had a share of approx. 57 % in Total cadmium emissions in Germany (compared to 32 % in 2000, 45 % in 2005 and about 57 % since 2008).
Targets	2020: 3.0 t 2030: 2.7 t 2050: 2.0 t
Assessment	 Value for the reference year (2005): 4.1 t [1] Latest available value (2015): 3.8 t [1] The target is a reduction of 22 % by 2020 compared to the reference value (the average value over the previous 5 years 2011–2015). The calculated trend based on the previous 5 years shows a reduction of 15 % by 2020. The relation between necessary change to achieve the target and the expected change due to the calculated trend is 34 %, which results in a yellow traffic light.

Comments Data	Considering the 'Protection of human health', besides the heavy metals mercury and lead especially cadmium, is relevant. If cadmium enters the human body in too high quantities, this element is toxic and mainly ingested with food. Ingestion with drinking water, beverages, through respiration (inhalation) or through the ingestion of soil or dust is insignificant in this respect. A suitable indicator for the health effects of cadmium would therefore be the amount this element taken in with food which is related to emissions from the energy sector. However, these values are not available and it would be very work-intensive to derive them from emission and dispersion models. Therefore those emissions were chosen as indicator which state the amount taken in with our food (higher emissions result in increased deposition on the soil surface, higher intake by plants and thus also with the food). Cadmium is one of the substances which plants can easily take in from the soil with their roots. The soil-plant transfer is especially significant with a low pH of the soil, but apart from that there can be large differences depending on the plant species. The pronounced soil-plant transfer therefore makes cadmium a 'critical' substance for the agricultural and horticultural use of the soil. Lead and other pollutants contained in sediment dust or contaminated soil material predominantly contaminate the surfaces of food crops. While this contamination can be almost completely removed during the preparation of the food, the cadmium concentration in vegetarian food cannot be reduced by respective measures. Cadmium can be also found in food of animal origin where it got via the food chain, for example through contaminated fodder plants or, as to fishes, shellfishes and mollusks, also through the water. Assuming an average consumption, the weekly cadmium intake amounts to almost 1.5 μg/kg body weight and thus to 58 % of the limit value TWI (Tolerable Weekly Intake). Frequent consumption results in an intake of 2.3 μg/kg body weight and week, irrespective of the sex, and thus equals a rate of 94 % of the TWI [2, 3].
Comments Targets	According to the 1998 Protocol on Heavy Metals, as amended on 13 December 2012 (ECE/EB.AIR/115) [4], each party shall reduce its total annual cadmium, mercury and lead emissions into the atmosphere compared to the emission level in the reference year by taking effective measures, appropriate to its particular circumstances. Two types of limit value are important for heavy metal emission control: • values for specific heavy metals or groups of heavy metals • values for emissions of particulate matter in general Basically, limit values for particulate matter cannot replace specific limit values for cadmium, lead and mercury because the quantity of metals associated with particulate emissions differs from one process to another. However, compliance with these limits contributes significantly to reducing heavy metal emissions in general. Moreover, monitoring particulate emissions in general is less expensive than monitoring the individual substances; also the continuous monitoring of individual heavy metals in general is not feasible. Therefore, particulate matter limit values are of great practical importance and are also laid down in the annex to the protocol, in most cases to complement specific limit values for cadmium, lead or mercury [4].

	For coal-fired power plants, which are the most relevant source for cadmium emissions, limit values for the emission of particulate matter are given in the annex to the protocol. Emissions therefore should be reduced by 26 % by the year 2020 and by 50 % by the year 2050 compared to the year 2005. These values will also be applied to cadmium. The value for the year 2030 is the result of a linear interpolation between the values for 2020 and 2050.
Literature	[1] UBA – Umweltbundesamt: Emissionen von Luftschadstoffen – Emissionsentwicklung 1990–2015 für Schwermetalle. 2016. https://www.umweltbundesamt.de/themen/luft/emissionen-von-luftschadstoffen, access 20 April 2017 [2] UBA – Umweltbundesamt: Aktualisierung der Stoffmonographie Cadmium – Referenz- und Human-Biomonitoring(HBM)-Werte. Stellungnahme der Kommission 'Human-Biomonitoring' des Umweltbundesamtes. Bundesge-sundheitsblatt, 8/2011. http://link.springer.com/article/10.1007%2Fs00103-011-1327-9#/page-1, access 13 November 2015 [3] BfR – Bundesinstitut für Risikobewertung, Fachgruppe Expositionsschätzung und -standardisierung, Abteilung Wissenschaftliche Querschnittsaufgaben: Aufnahme von Umweltkontaminanten über Lebensmittel (Cadmium, Blei, Quecksilber, Dioxine und PCB). Ergebnisse des Forschungsprojektes LExUKon, 2010. http://www.bfr.bund.de/cm/350/aufnahme_von_umweltkontaminanten_ueber_lebensmittel.pdf, access 13 November 2015 [4] UNECE – United Nations Economic Commission for Europe, Executive Body for the Convention on Long-range Transboundary Air Pollution (ECE/EB.AIR/115): 1998 Protocol on Heavy Metals, as amended on 13 December 2012. 2014. http://www.unece.org/fileadmin/DAM/env/documents/2012/EB/ECE.EB.AIR.115_ENG.pdf, access 13 November 2015

3. Energy-related Emissions of Mercury

Justification referring to Sustainability Concept	This indicator addresses the sustainability rule *'Protection of human health'*: Protecting every individual's life and health is an existential issue of sustainability based on the precautionary principle. The aim is to avoid health problems and risks actually or potentially caused by the energy system, e.g., noise, emissions of air pollutants and particulate matter, radioactive radiation, etc.
Description	Mercury is toxic if it enters the human body in too high quantities and is mainly ingested with the food.
Unit	t/a

Data

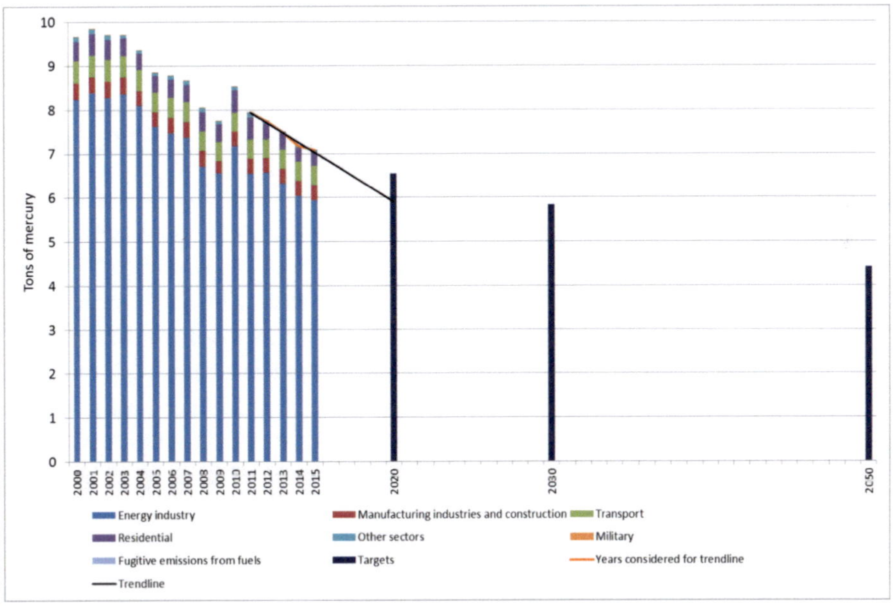

Figure 1: Energy-related emissions of mercury (own diagram based on [1])

In 2015, energy-related emissions of mercury had a share of approx. 78 % in total mercury emissions in Germany (this share steadily increased from 54 % in 2000 over 64 % in 2005).

Targets	2020: 6.5 t 2030: 5.8 t 2050: 4.4 t
Assessment	Value for the reference year (2005): 8.7 t [1] Latest available value (2015): 7.1 t [1] The target is a reduction of 13 % by 2020 compared to the reference value (the average value over the previous 5 years from 2011-2015). The calculated trend is the expected change and results in a reduction of 21 %. The expected over-achievement of the target of 2020 is assigned with a green traffic light.
Comments Data	Considering the 'Protection of human health', especially the heavy metals cadmium, mercury, and lead are relevant. If they enter the human body in too high quantities, these elements are all toxic, although in different ways. In addition, the three of them are all

	mainly ingested with the food. Ingestion with drinking water, beverages, through respiration (inhalation) or through the ingestion of soil or dust is insignificant in this respect. A suitable indicator for the health effects of heavy metals would therefore be the amount of heavy metals taken in with the food, which is related to emissions from the energy sector. However, these values are not available and it would be very work-intensive to derive them from emission and dispersion models. Therefore those emissions were chosen as indicator which state the amount taken in with our food (higher emissions result in increased deposition on the soil surface, higher intake by plants and thus also with our food). The toxicological reference value for mercury is the PTWI (Provisional Tolerable Weekly Intake) of 1.6 µg/kg body weight per week for methylmercury, which was determined by the JECFA (Joint FAO/WHO Expert Committee on Food Additives) and confirmed by the EFSA (European Food Safety Authority) in 2004; in the U.S. this value is 0.10 µg/(kg*d) [2] (which corresponds to 0.7 µg/kg body weight per week). For the toxicological classification of total mercury an assessment value of 2.4 µg/kg body weight per week was stated which takes account of the JECFA's request that the share of methylmercury in total mercury must not exceed two thirds. A weekly intake of total mercury of 0.49 µg/kg body weight for the average con-sumer of the general population corresponds to 21 % of the assessment value. Frequent consumption results in a share of approx. 37 %. A weekly intake of methylmercury of 0.17 µg/kg body weight for the average consumer of the gen-eral population corresponds to 11 % of the PTWI of JECFA. Frequent consumption results in a share of approx. 34 % [2]. If the US value is taken as a basis (0.7 µg/kg body weight per week), this corresponds to 25 % of the PTWI for an average consumer; frequent consumption results in a share of approx. 78 %. These values already indicate that mercury emissions have to be reduced significantly. Emission limit for mercury in coal-fired power plants (see [3]): The German emission limit for mercury in coal-fired power plants is a daily mean of 30 µg/Nm3 and (after 2019 for old installations) 10 µg/Nm3 as an annual mean. In the US, the legislature took action in 2012 due to the described health risks and the resulting imminent economic losses and radically lowered the permitted emission limits for mercury in coal-fired power plants. These American limit values for power plants are – converted into the units used in Germany – 1.4 µg/Nm3 for anthracite-fired and 4.1 µg/Nm3 for lignite-fired power plants (respective mean values for 30 days). They are considerably lower than the above-mentioned German limit values. According to the Minamata Convention on Mercury, which was also signed by Germany [4], the best available technology has to be used for new coal-fired power plants for the protection against mercury emissions. Since the above-mentioned American limit values are reached in practice, they should also apply for new German coal-fired power plants (see also [5]).
Comments Targets	According to the 1998 Protocol on Heavy Metals, as amended on 13 December 2012 (ECE/EB.AIR/115) [6], each party shall reduce its total annual cadmium, mercury and lead emissions into the atmosphere compared to the emission level in the reference year by taking effective measures, appropriate to its particular circumstances.

	Two types of limit value are important for heavy metal emission control: • Values for specific heavy metals or groups of heavy metals and • Values for emissions of particulate matter in general. Basically, limit values for particulate matter cannot replace specific limit values for cadmium, lead and mercury because the quantity of metals associated with particulate emissions differs from one process to another. However, compliance with these limits contributes significantly to reducing heavy metal emissions in general. Moreover, monitoring particulate emissions in general is less expensive than monitoring the individual metals; also the continuous monitoring of individual heavy metals in general is not feasible. Therefore, particulate matter limit values are of great practical importance and are also laid down in the annex to the protocol, in most cases to complement specific limit values for cadmium, lead or mercury [6]. For coal-fired power plants, which are the most relevant source for mercury emissions, limit values for the emission of particulate matter are given in the annex to the protocol. Emissions therefore should be reduced by 26 % by the year 2020 and by 50 % by the year 2050. These values will also be applied to mercury. The value for the year 2030 is the result of a linear interpolation between the values for 2020 and 2050.
Literature	[1] UBA – Umweltbundesamt: Emissionen von Luftschadstoffen – Emissionsentwicklung 1990–2015 für Schwermetalle, 2016. http://www.umweltbundesamt.de/themen/luft/emissionen-von-luftschadstoffen, access 20 April 2017 [2] BfR, Fachgruppe Expositionsschätzung und -standardisierung, Abteilung Wissenschaftliche Querschnittsaufgaben: Aufnahme von Umweltkontaminanten über Lebensmittel (Cadmium, Blei, Quecksilber, Dioxine und PCB). Ergebnisse des Forschungsprojektes LExUKon, 2010. http://www.bfr.bund.de/cm/350/aufnahme_von_umweltkontaminanten_ueber_lebensmittel.pdf, access 13 November 2015 [3] Zeschmar-Lahl, B.: Quecksilberemissionen aus Kohlekraftwerken in Deutschland – Stand der Technik der Emissionsminderung. http://www.bzl-gmbh.de/de/sites/default/files/BZL_Studie_QuecksilberemissionenAusKohlekraftwerkenInDeutschland_final%281%29.pdf, access 13 November 2015 [4] UNEP – United Nations Environment Programme: Minamata Convention on Mercury. October 2013. http://www.mercuryconvention.org/Portals/11/documents/Booklets/Minamata%20Convention%20on%20Mercury_booklet_English.pdf, access 26 October 2015 [5] Klein, M.: Sind starre Emissionsgrenzwerte das richtige Mittel, den Eintrag von Schwermetallen in die Umwelt nachhaltig zu senken? Betrachtungen am Beispiel Quecksilber. In: Beckmann, M.; Hurtado, A.: Kraftwerkstechnik 2015. Strategien, Anlagentechnik und Betrieb. Freiberg, 2015, pp. 97–605 [6] UNECE – United Nations Economic Commission for Europe, Executive Body for the Convention on Long-range Transboundary Air Pollution (ECE/EB.AIR/115): 1998 Protocol on Heavy Metals, as amended on 13 December 2012. 2014. http://www.unece.org/fileadmin/DAM/env/documents/2012/EB/ECE.EB.AIR.115_ENG.pdf, access 13 November 2015

4. Energy Import Dependency

<table>
<tr><td>Justification referring to Sustainability Concept</td><td>This indicator addresses the sustainability rule 'Satisfaction of basic needs': For all members of society, a minimum of basic services and protection against key risks to life must be guaranteed. Access to sufficient energy or energy services to live a decent life is certainly among these basic needs. The necessary amount of energy has to be secured to ensure a life in dignity and justice for everybody.</td></tr>
<tr><td>Description</td><td>Import dependency measures the extent to which Germany relies on imports to meet its energy requirements. It is defined as the ratio of energy resource imports (in PJ) to total primary energy supply (the value of primary energy consumption) per year for Germany (in PJ). The calculation of the primary energy consumption is based on the efficiency principal defined by the AG Energiebilanzen since 1995 [1]. For fuels like coal, lignite and natural gas a calorific value is assumed and multiplied with the amount of used fuel in order to calculate the primary energy. For the generation of electricity by means of renewables like wind, hydro power and photovoltaic the primary energy consumption is defined by the total amount of produced electricity. These means an efficiency of 100 % is assumed. For electricity generation by nuclear power plants an efficiency of 33 % is assumed. Thus, a reduction or substitution of nuclear power supply by renewables of wind and photovoltaic power leads to a reduction of both the amount of energy resource imports and the primary energy consumption.</td></tr>
<tr><td>Unit</td><td>%</td></tr>
<tr><td>Data</td><td>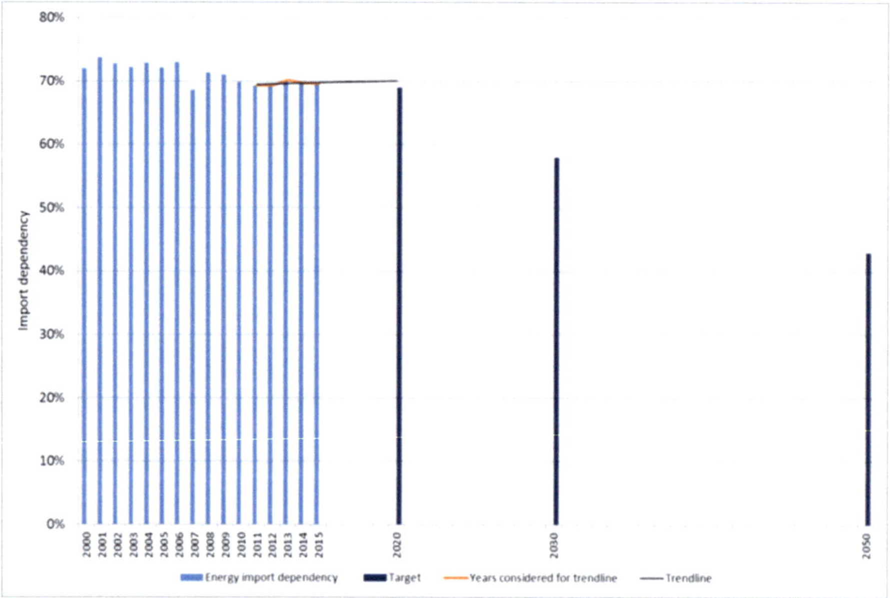

Figure 1: Energy import dependency (net imports in % of primary energy consumption) (own diagram based on [2])</td></tr>
<tr><td>Targets</td><td>2020: 69 %
2030: 58 %
2050: 43 %</td></tr>
</table>

Assessment		Latest available value (2015): 70 % The trend calculated based on the previous 5 years (2011–2015) shows an increase in the share of imported energy of 0.7 % by 2020 compared to thereference value (the average value over the years 2011 to 2015). The necessary change (the relation between target of 2020 and the reference value) requires a reduction of 0.8 %. This is a deviation of 187 %, which results in a red traffic light.
Comments Data	For the different energy carriers, the following import dependencies are given in [2]: • Hard coal: steadily increasing from 7.7 % in 1990 to 44.8 % in 2000 and 88.5 % in 2015 • Lignite: small amounts exported in all years • Mineral oil: more than 95 % imported in all years • Natural gas: slightly increasing from 75 % in 1990 to 79.1 % in 2000 and 81.3 % in 2010 • Nuclear fuel elements: 100 % imported in all years • Renewable energies are not considered as imported here Beside the total amount of imported energy ressources nother important aspect of a country's import dependency is the diversification of the supplying countries. Russia, for example, is the dominant supplier of three main fuel types. In terms of crude oil imports in 2013, the largest shares came from Russia (41 %), Norway (14 %), Great Britain (13 %), Nigeria (9 %), Kazakhstan (9 %), and Lybia (8 %) [4]. For natural gas, the top suppliers in 2013 were Russia (38 %), the Netherlands (26 %), Norway (20 %), and Denmark (6 %) [5]. Hard coal imports in 2013 were dominated by Russia (29 %), the USA (25 %), Colombia (19 %), EU countries (12 %), South Africa (7 %), and Australia (4 %) [6]. To get a clearer picture of the reliability of the import partner countries, one might include the countries' specific (political) risks. To operationalize this, the coded level of democracy in these countries can be taken into consideration, e.g., by using the Polity IV index developed and applied by political science projects [7]. Another way to operationalize import dependency is proposed in [8. p. Z-18]. 'An appropriate indicator can be derived from the calculation of relative market shares. We shall first compare the market share of the German sales market from the perspective of an exporting country (e.g., Russia) with the import share of this exporting country from the perspective of Germany. The larger the ratio is, the less critical is the supply risk for Germany regarding the imports from the corresponding supplier country'.	
Comments Targets	In [3, p.10] data for the use of different types of energy carriers are given for the years 2010, 2020, 2030, 2040, and 2050. Here, it is assumed that throughout all years the import shares of the different fuel types will remain the same as the current ones: • Mineral oil: 99.5 % • Hard coal: 88.5 % • Gas: 88.9 % • Lignite: 0 % • Nuclear fuel elements: 0 % (after 2022) Furthermore, it is assumed that the share of used coal – lignite and hard coal – will be the same as in 2015. As a result, the target values are as above.	

Literature	[1] Ziesing, H.-J., Görgen, R., Maaßen U., Nickel, M. et al.: Energie in Zahlen. Arbeit und Leistungen der AG Energiebilanzen. WAZ-Druck, Duisburg, 1. Oktober 2012. http://www.ag-energiebilanzen.de/index.php?article_id=34&clang=0 [2] BMWi – Bundesministerium für Wirtschaft und Energie: Gesamtausgabe der Energiedaten – Datensammlung des BMWi. Letzte Aktualisierung: 30.01.2017. http://www.bmwi.de/Redaktion/DE/Artikel/Energie/energiedaten-gesamtausgabe.html, access 30 May 2017 [3] DLR – Deutsches Zentrum für Luft- und Raumfahrt; IfNE – Ingenieurbüro für neue Energien; IWES – Fraunhofer Institut für Windenergie und Energiesystemtechnik: Langfristszenarien und Strategien für den Ausbau der erneuerbaren Energien in Deutschland bei Berücksichtigung der Entwicklung in Europa und global. Schlussbericht BMU – FKZ 03MAP146, March 2012 [4] http://www.dlr.de/dlr/Portaldata/1/Resources/bilder/portal/portal_2012_1/leitstudie2011_bf.pdf, access 14 June 2016 [5] MWV – Mineralölwirtschaftsverband e. V.: Jahresbericht 2013/Mineralöl-Zahlen. July 2014. [6] https://www.mwv.de/wp-content/uploads/2016/06/mwv-publikationen-jahresberichte-mineraloelzahlen-2013.pdf, access 16 June 2016 [7] BDEW – Bundesverband der Energie- und Wasserwirtschaft e. V.: Energie-Info. Entwicklung der Energieversorgung 2014. [8] https://www.bdew.de/internet.nsf/id/0CDCAA69F51E56CFC1257E0C0047C7E3/$file/Entwicklung_der_Energieversorgung_2014.pdf, access 15 June 2016 [9] Statistisches Bundesamt: DESTATIS Genesis Datenbank: Erhebung über die Einfuhr von Kohle. Wiesbaden, 2014 [10] Polity IV Project. http://www.systemicpeace.org/polityproject.html, access 25 January 2016 [11] Löschel, A.; Erdmann, G; Staiß, F.; Ziesing, H.: Expertenkommission zum Monitoring-Prozess 'Energie der Zukunft'. Stellungnahme zum zweiten Monitoring-Bericht der Bundesregierung für das Berichtsjahr 2012. Berlin, March 2014 https://www.bmwi.de/Redaktion/DE/Downloads/M-O/monitoringbericht-energie-der-zukunft-stellungnahme.html, access 15 June 2016

5. Monthly Energy Expenditures of Households with a Monthly Net Income less than 1,300 Euro

<table>
<tr><td>Justification referring to Sustainability Concept</td><td>This indicator addresses the sustainability rule 'Satisfaction of basic needs': For all members of society, a minimum of basic services and protection against key risks in life must be guaranteed. Access to sufficient energy or energy services to live a decent life is certainly among these basic needs. The necessary amount of energy has to be secured to ensure a life in dignity and justice for everybody.</td></tr>
<tr><td>Description</td><td>Energy expenditures of German households with low income (monthly income less than 1,300 €) are reported here.</td></tr>
<tr><td>Unit</td><td>€/month</td></tr>
<tr><td>Data</td><td>
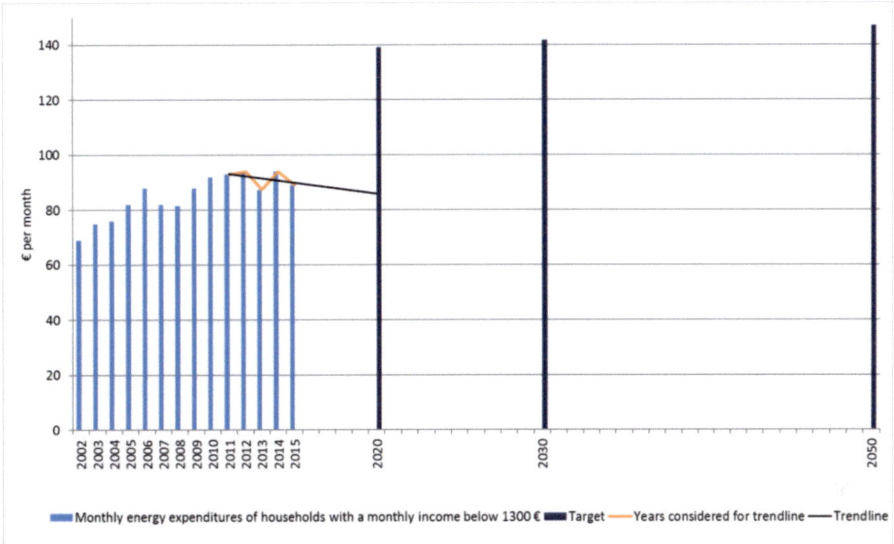

Figure 1: Monthly energy expenditures of households with a net income below 1,300 € (own diagram based on [1-5])</td></tr>
<tr><td>Targets</td><td>2020: 139 €/month
2030: 142 €/month
2050: 147 €/month</td></tr>
<tr><td>Assessment</td><td>

Latest available value (2015): 89 €/month [1]

On average over the period 2011 to 2015, expenditures of households with a net household income below 1,300 € amounted to 89 € per month. Since the trendline shows a decreasing monthly expenditure not reaching the maximum target value for 2020, a green traffic light was assigned to this indicator.</td></tr>
<tr><td>Comments Data</td><td>Expenditures of low-income households have been vividly discussed from different sides and are often associated with a definition and measurement of 'energy poverty', 'fuel poverty' or 'electricity poverty' respectively. However, there is little agreement on the problem definition and the choice of mathematical measures. It has also been proven that a quantitative assessment of 'fuel poverty' or similarly defined problems depends heavily</td></tr>
</table>

	on the measuring method [6, p. 32f]. This bias should be avoided, and the measuring method and its assumptions should be transparent. Hence, we propose not to include expenditures of low-income households for energy supply a priori into a problem definition but to first statistically raise relevant expenditures in a detailed manner by statistical offices. We refrain from an evaluation of the data without mathematical methods as outlined, e.g., in [7]. We rather suggest an evaluation focusing on the expenditures normally required to meet the basic energy needs: Statistically raised data about expenditures should be compared to these 'essential expenditures'. This follows the proposal of Kerstin Tews [7] and is aimed at quantifying the expenditures generally required for an adequate energy supply. 'Essential expenditures' in Germany still need to be defined for different household types. In the United Kingdom, a model called BREDEM, which is based on data of the English Household Survey, is applied to identify these expenditures. Moreover, the number of persons/households should be surveyed that permanently do not spend more than the essential expenditures because they choose to save money for other important needs of their household. This is important because that particular group would not be identified as suffering from a high financial burden of energy bills. However, so far no data are available to assess 'energy poverty' in the way described. It is therefore proposed to draw on 'monthly energy expenditures of households with a monthly net income less than 1,300 euros.' Monthly net income of households is categorized according to the statistics of the Federal Statistical Office. According to the Statistical Office, net household income is calculated by subtracting income/wage tax, church tax, and solidarity surcharge as well as the mandatory social security contributions from gross household income (total income of the household from employment, property, public and private transfers, and subletting). According to [3], households with an income between 3,600 € and 5,000 € and between 5,000 € and 18,000 € together constitute about 46 % of all households in 2012. Their monthly expenditure for energy is much higher than for the other income groups. Thus, an energy saving program designed for the highest income groups is most promising. Data for energy expenses from the Federal Statistical Office include: • electricity • fees • fuel costs (for heating) • taxes or levies on heating plants Data for 2002 to 2012 have been taken from [2]. In [2], data on monthly energy expenditures are given directly for the income class below 1,300 €. Data for 2013 have been calculated from information given in [4] and are the weighted average of the income classes below 500 € (2.57 % of households), between 500 and 900 € (39.5 % of households), and between 900 and 1,300 € (57.9 % of households). Data for 2014 and 2015 are taken from [5] and [1], since data for the income class below 1,300 € were resumed.
Comments Targets	According to [8], the expenditure of low-income households for heating should not exceed 10 % of income (fuel poverty). However, the mentioned source is from 1991 and is based on data from the UK. The data from [8] are also addressed and critically discussed in [9]. However, since no further data are available (current figures and for German standards), the 10 % share is used to determine the target. According to [10], German households

	spend on average 70 % of their energy expenditure on heating and 30 % on electricity. With these values, reasonable energy expenditures of about 15 % of net disposable household income are obtained. Households with an income below 1,300 € had a net household income of 901 € in the year 2011 [2] and 916 € in the year 2015 [1].
Literature	[1] Destatis (Statistisches Bundesamt): Laufende Wirtschaftsrechnungen - Einkommen, Einnahmen und Ausgaben privater Haushalte 2015, Fachserie 15 Reihe 1. [2] https://www.destatis.de/DE/Publikationen/Thematisch/EinkommenKonsum-Lebensbedingungen/EinkommenVerbrauch/EinnahmenAusgabenprivaterHaushalte2150100157004.pdf?__blob=publicationFile, access 25 January 2017 [3] Destatis (Statistisches Bundesamt): Laufende Wirtschaftsrechnungen – Einkommen, Einnahmen und Ausgaben privater Haushalte 2002–2012, Fachserie 15 Reihe 1. https://www.destatis.de/GPStatistik/receive/DESerie_serie_00000152, access 20 October 2016 [4] Destatis (Statistisches Bundesamt): Wirtschaftsrechnungen: Einkommens- und Verbrauchsstichprobe – Aufwendungen privater Haushalte für den Privaten Konsum –2008. Fachserie 15 Heft 5 https://www.destatis.de/DE/Publikationen/Thematisch/EinkommenKonsumLebensbedingungen/Konsumausgaben/EVS_AufwendungprivaterHaushalte2152605089004.pdf?__blob=publicationFile, access 20 October 2016 [5] Destatis (Statistisches Bundesamt): Wirtschaftsrechnungen: Einkommens- und Verbrauchsstichprobe – Einnahmen und Ausgaben privater Haushalte für den Privaten Konsum – 2013. Fachserie 15 Heft 5, 2015. https://www.destatis.de/DE/Publikationen/Thematisch/EinkommenKonsumLebensbedingungen/Konsumausgaben/EVS_AufwendungprivaterHaushalte2152605139004.pdf?__blob=publicationFile, access 20 October 2016 [6] Destatis (Statistisches Bundesamt): Laufende Wirtschaftsrechnungen – Einkommen, Einnahmen und Ausgaben privater Haushalte 2014, Fachserie 15 Reihe 1. https://www.destatis.de/DE/Publikationen/Thematisch/EinkommenKonsum-Lebensbedingungen/EinkommenVerbrauch/EinnahmenAusgabenprivaterHaushalte2150100147004.pdf?__blob=publicationFile, access 25 January 2017 [7] Heindl, P.: Measuring Fuel Poverty: General Considerations and Application to German Household Data. Discussion Paper No. 13-046, 2013 [8] Tews, K.: Energiearmut definieren, identifizieren und bekämpfen. FFU-Report 04-2013, 2013 [9] Boardman, B.: Fuel Poverty: From Cold Homes to Affordable Warmth. Belhaven, London, 1991 [10] Pye, S.; Dobbins, A.: Energy poverty and vulnerable consumers in the energy sector across the EU: analysis of policies and measures. Insight_E, May 2015 [11] Umweltbundesamt: Private Haushalte und Konsum. https://www.umweltbundesamt.de/daten/private-haushalte-konsum/energieverbrauch-privater-haushalte, access 20 October 2016

6. SAIDI of Electricity

Justification referring to Sustainability Concept	This indicator addresses the sustainability rule *'Satisfaction of basic needs'*: For all members of society, a minimum of basic services and protection against key risks to life must be guaranteed. The access to sufficient energy or energy services to live a decent life is certainly among these basic needs. The necessary amount of energy has to be secured to ensure a life in dignity and justice for everybody. For certain industry customers who depend on non-interrupted electricity, the supply reliability and quality is also a basic need.
Description	The average supply interruption in minutes per year per connected consumer is reported for the electricity grid. The SAIDI (System Average Interruption Duration Index) of electricity is calculated separately as an average for all affected consumers in medium- and low-voltage grids and then summed up. It includes non-scheduled interruptions due to atmospheric impacts, influence of third parties, missed responsibilities of the network operator, and retroactive disturbance from other grids. Only interruptions longer than three minutes are considered.
Unit	min/a
Data	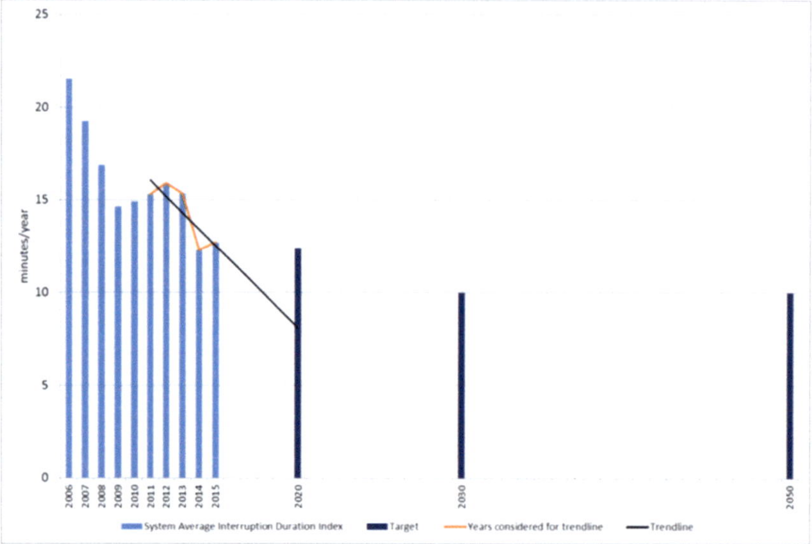 Figure 1: System Average Interruption Duration Index of electricity 2006-2015 (own diagram based on [1])
Targets	2020: 12.5 min 2030: 10.0 min 2050: 10.0 min
Assessment	 Latest available value (2015): 12.7 min/a [1] The target is a value of 12.5 minutes for 2020. The trend based on the previous 5 years (2011–2015) shows lower values than the given maximum target. Thus, a green traffic light is assigned. However, this outcome is a result of the relatively low SAIDI values for 2014 and 2015. If the trend have been calculated based on the data for the years 2009 to 2013, the trend would show increasing values; the targets would by far not be fulfilled and a red traffic light would be assigned.

Comments Data	In general, the German grid performs very well in international comparisons: It is amongst those with the least SAIDI in Europe. However, the short-term security of supply is less guaranteed, for example. Therefore, different experts suggested to statistically determine the number of all interruptions (regardless of their duration) and to find an index, which reflects the quality of voltage. In addition, the expert commission [2] strongly recommends a measure that also takes planned supply disruptions and interruptions of less than three minutes into account. Interruptions of less than 3 minutes duration can lead to impairments and probably economic losses for electricity customers with sensitive industrial processes. Short supply interruptions (less than 1 second) accounted for the vast majority of interruptions (59 % in the period 2006–2008 and 72 % in the period 2009–2011 [3]), whereas interruptions between ten seconds and three minutes quadrupled within a few years. Furthermore, it is important to differentiate between reliability of supply and quality of voltage. IT devices that are very sensitive to voltage drops have spread quickly in Germany. Thus, it is also important to monitor the voltage quality in addition to interruptions as supply quality and reliability are decisive location factors for businesses [4]. Finally, the reliability of supply differs depending on the geographic location of the grid customers in the supply area.
Comments Targets	There is no official target value for SAIDI in Germany. A viable comparison is possible with values for EU's top ranking countries in 2013: Denmark had 11.25 min, Luxemburg reported 10 min for the year 2013 [5]. Based on the data of Luxemburg, a target value of 10 min was chosen for 2030 and for 2050. The target for 2020 is a result of an interpolation between the average value for the previous 5 years and the chosen value for 2030.
Literature	[1] Bundesnetzagentur: Versorgungsqualität – SAIDI-Werte 2006–2015. Oktober 2016. https://www.bundesnetzagentur.de/cln_1421/DE/Sachgebiete/ElektrizitaetundGas/Unternehmen_Institutionen/Versorgungssicherheit/Stromnetze/Versorgungsqualitaet/Versorgungsqualitaet-node.html [2] Löschel, A.; Erdmann, G; Staiß, F.; Ziesing, H. - Expertenkommission zum Monitoring-Prozess 'Energie der Zukunft': Stellungnahme zum zweiten Monitoring-Bericht der Bundesregierung für das Berichtsjahr 2012. 2014 [3] Verteilung der Störungen bei der Stromversorgung von Industriekunden nach Art der Vorfälle in Deutschland im Zeitraum von 2006 bis 2008 und 2009 bis 2011. http://de.statista.com/statistik/daten/studie/214716/umfrage/stromversorgungsqualitaet-im-industriekundenbereich/, access 20 October 2016 [4] Von Roon, S., Forschungsgesellschaft für Energiewirtschaft mbH: Versorgungsqualität und -zuverlässigkeit als Standortfaktor. 2013 https://www.ffegmbh.de/download/veroeffentlichungen/351_tagung2013_roon/Tagung2013_roon.pdf, access 20 October 2016 [5] Council of European Energy Regulators: CEER Benchmarking Report 5.2 on the Continuity of Electricity Supply. Data update, 12 February 2015.

7. Relation of Employees in the Renewable Energy Sector to Total Employees

Justification referring to Sustainability Concept	This indicator addresses the sustainability rule *'Autonomous subsistence based on income from own work'*: For all members of society, the possibility of ensuring their livelihood through freely acquired occupation has to be assured. Concerning the energy system, this includes income generation possibilities within the energy sector.
Description	The indicator reports the relation of employees within the Renewable Energy (RE) sector to the total number of employees.
Unit	%
Data	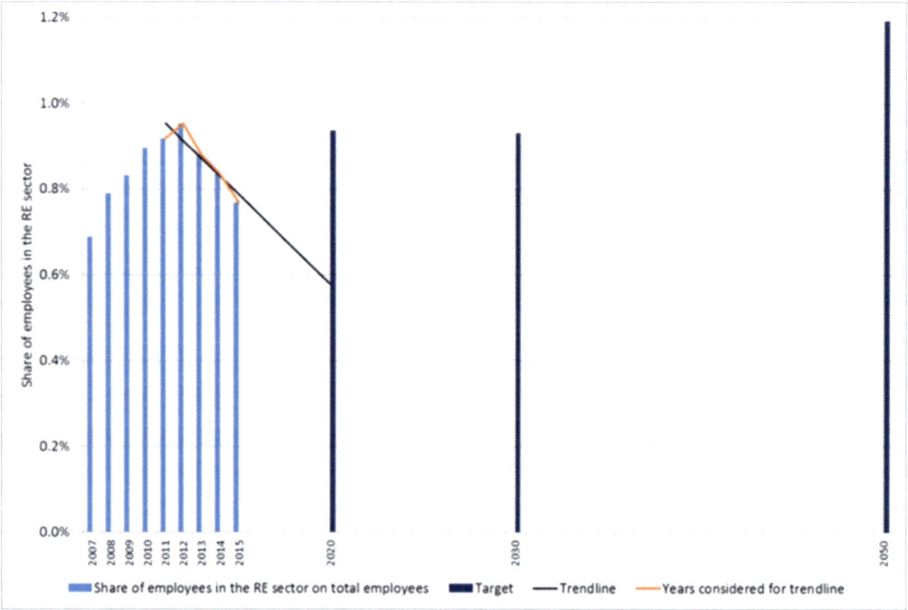 Figure 1: Share of employees in the RE sector in relation to the total number of employees (own diagram based on [1-3])
Targets	2020: 0.94 % 2030: 0.93 % 2050: 1.19 %
Assessment	 Latest available value (2015): 0.77 % [1] The target is a relation of 0.94 % for 2020, which is about 8 % higher than the reference value (as a consequence of the fact that investments in 2020 will be lower [4]). However, the trend shows a decrease of about 34 % in 2020 resulting in a red traffic light.
Comments Data	The number of employees in the RE sector steadily increased from 277,300 in 2007 to 399,800 in 2012. Then the number decreased to 355,000 in 2014 and 330,000 in 2015 [1-3].

	The latest available value for 2015 of the share of employees in the RE sector was calculated based on data for employees in the REsector given in [1, p. 8, investments given in 3] and on data for total employees, given in [5]). This indicator addresses the relation of employees in the RE sector to total em-ployees. It also includes employment in the production of exported RE compounds. However, since the sustainability analysis of the energy transition concentrates on Germany, employment in other countries is not taken into consideration (e.g. a high share of PV modules is produced in China). Also possible cutbacks in employment in the conventional energy sector and other sectors as a consequence of higher energy costs resulting from subsidies for RE are not considered; it is difficult to generate these data and will only be possible based on models which are subject to several assumptions.
Comments Targets	The number of employees depends on • the volume of investments in Germany • the export of RE technologies • maintenance and operation of plants. • Federal research expenditures on renewable energy In [4] information is given on the volume of investments in Germany until 2050. Data on future exports are not available from literature and can therefore not be considered. An assessment of the share of employees who are responsible for maintenance and operation of the plants is also not possible with the data available from literature. Therefore, the number of employees is only estimated here based on the respective volume of investments. In 2015, the volume of investments in the field of RE amounted to 15 billion euros, the number of employees to 330,000. The yearly volume of investments given in [4] accounted for 18.4 billion euros until 2020, 17.2 billion euros until 2030, 18.7 billion euros until 2040 and 19.9 billion euros until 2050. Based on these values, we can calculate: • 416,000 employees for 2020 • 387,000 employees for 2030 • 449,000 employees for 2050 [4] states that ‘perspectively, an increase of gross employment to 530,000 to 640,000 people in 2030 is possible’. However, this number considers the fact that ‘the global technological leadership of the German industry in the field of RE which was brought about by the Renewable Energy Law should also lead to a considerable advantage on the growing future world energy market’. The indicator shows the share in the total employment. 42,979 million employees were assumed for 2015, in 2011 this figure was 41,523 million [5]. Starting with the average value (0.87 %) over the previous 5 years (2011-2015) for the ‘Relation of Employees in the RE Sector’ the following target values using the data given in [4] for the investments and the total number of employees can be calculated: • 2020: 38.6 million employees in total and 361925 employees in RE: 0.94 % • 2030: 36.2 million employees in total and 336989 employees in RE: 0.93 % • 2050: 32.8 million employees in total and 391004 employees in RE: 1.2 %

Literature	[1] GWS, Fraunhofer ISI, DIW Berlin, DLR, Prognos, ZSW. Bruttobeschäftigung durch erneuerbare Energien in Deutschland und verringerte fossile Brennstoff-importe durch erneuerbare Energien und Energieeffizienz. Stand September 2016. [2] https://www.bmwi.de/Redaktion/DE/Downloads/S-T/bruttobeschaeftigung-erneuerbare-energien-monitioringbericht-2015.pdf?__blob=publicationFile&v=11, access 12 April 2017 [3] GWS, DIW, DLR, Prognos, ZSW (2015): Beschäftigung durch erneuerbare Energien in Deutschland: Ausbau und Betrieb, heute und morgen. Studie im Auftrag des Bundesministeriums für Wirtschaft und Energie. Osnabrück, Berlin, Stuttgart. [4] BMWi – Bundesministerium für Wirtschaft und Energie: Fünfter Monitoringbericht zur Energiewende. Die Energie der Zukunft. Berichtsjahr 2015. https://www.bmwi.de/Redaktion/DE/Publikationen/Energie/fuenfter-monitoring-bericht-energie-der-zukunft.pdf?__blob=publicationFile&v=34, access 12 April 2017 [5] DLR – Deutsches Zentrum für Luft- und Raumfahrt; IfNE – Ingenieurbüro für neue Energien; IWES – Fraunhofer Institut für Windenergie und Energiesystemtechnik: Langfristszenarien und Strategien für den Ausbau der erneuerbaren Energien in Deutschland bei Berücksichtigung der Entwicklung in Europa und global. Schlussbericht BMU, FKZ 03MAP146, March 2012 [6] http://www.dlr.de/dlr/Portaldata/1/Resources/bilder/portal/portal_2012_1/leitstudie2011_bf.pdf, access 12 April 2017 [7] Destatis: Number of employees 2006 to 2015. https://www.destatis.de/DE/ZahlenFakten/Indikatoren/Konjunkturindikatoren/Arbeitsmarkt/karb811.html, access 12 April 2017

8. Final Energy Consumption of Private Households per Capita

<table>
<tr><td>Justification referring to Sustainability Concept</td><td>This indicator addresses the sustainability rule 'Just distribution of opportunities to use natural resources': Usage of the environment by the energy system has to be justified and allocated under fair participation of all stakeholders. This refers both to intergenerational (time) and intragenerational (spatial) allocation, the attainment of latter without compromising the intergenerational rights of usage. It also implies the fair allocation of responsibilities and burdens among all stakeholders.and</td></tr>
<tr><td>Description</td><td>The final energy consumption of private households per capita is reported here. This includes heat and electricity, but not mobility.</td></tr>
<tr><td>Unit</td><td>GJ per capita</td></tr>
<tr><td>Data</td><td>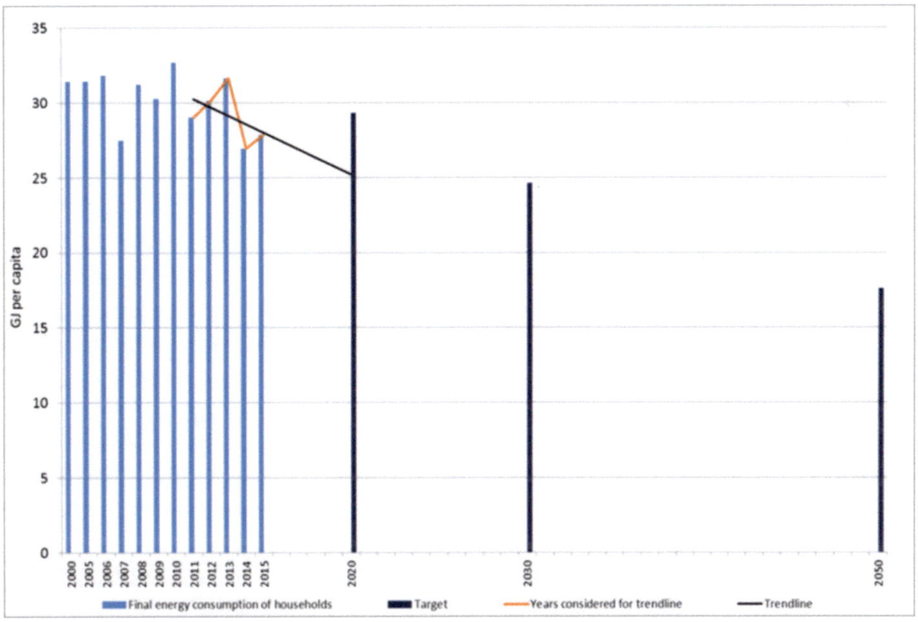

Figure 1: Final energy consumption in private households (own diagram based on [1])
(* for 2011 to 2015 data are based on the follow-up of the results of the population census in 2011, due to this statistical effect the population from Germany is around 1,5 million reduced from the year 2011 onwards than before).</td></tr>
<tr><td>Targets</td><td>2020: 29.3 GJ per capita
2030: 24.7 GJ per capita
2050: 17.6 GJ per capita</td></tr>
</table>

Assessment	 Value for the reference year (2008): 31.2 GJ per capita [1] Latest available value (2015): 27.9 GJ per capita [1] Based on the previous 5 years (2011-2015), the calculated trend shows a decrease of the final energy consumption in private households per capita by 13.3 % in 2020 compared to the reference value (the average value over the years 2011 to 2015). The target for 2020 requires a decrease by only 1 % compared to the reference value. Therefore, so far this indicator is on track to reach the target for the year 2020 and gets a green traffic light.
Comments Data	Data for final energy consumption in households are taken from [1] and are based on data given in Arbeitsgemeinschaft Energiebilanzen. Population data for Germany are taken from [2]. Figure 1 shows the specific final energy consumption of households (GJ per capita). If this is used to calculate the total final energy consumption in the household sector, this results in approx. 2,560 PJ for 2008. This value corresponds very well with the data given in [3].
Comments Targets	Final energy consumption of households should be reduced by the same share as total primary energy consumption (reduction of 20 % until 2020, of 30 % until 2030 and of 50 % until 2050 compared to 2008, see [4]). With a population of 82 million in 2008, final energy consumption of households amounted to 2,560 PJ. This results in a target value of 2,046 PJ for 2020 (20 % reduction), of 1,791 PJ for 2030 (30 % reduction) and 1,279 PJ for 2050 (50 % reduction). With a population of 80.5 million in 2020, of 79.1 million in 2030 and of 73.8 million in 2050 (prognosis of population development given in [3]), and based on the assumption of a declining population, this results in the mentioned target values for final energy consumption of households per capita. Due to the fact that a reduction of primary energy consumption is easier to achieve than a reduction of final energy consumption, the chosen target values are more ambitious.
Literature	[1] UBA – Umweltbundesamt: Energieverbrauch privater Haushalte. https://www.umweltbundesamt.de/daten/private-haushalte-konsum/energieverbrauch-privater-haushalte, access 10 April 2017. [2] Statista. Einwohnerzahl - Anzahl der Einwohner von Deutschland von 1990 bis 2015 (in Millionen). https://de.statista.com/statistik/daten/studie/2861/umfrage/entwicklung-der-gesamtbevoelkerung-deutschlands/, access 10 April 2017 [3] DLR – Deutsches Zentrum für Luft- und Raumfahrt; IfNE – Ingenieurbüro für neue Energien; IWES – Fraunhofer Institut für Windenergie und Energiesystemtechnik: Langfristszenarien und Strategien für den Ausbau der erneuerbaren Energien in Deutschland bei Berücksichtigung der Entwicklung in Europa und global. Schlussbericht BMU, FKZ 03MAP146, March 2012 http://www.dlr.de/dlr/Portaldata/1/Resources/bilder/portal/portal_2012_1/leitstudie2011_bf.pdf, access 25 May 2015 [4] BMWi – Bundesministerium für Wirtschaft und Technologie; BMU – Bundesministerium für Umwelt, Naturschutz und Reaktorsicherheit: Energiekonzept für eine umweltschonende, zuverlässige und bezahlbare Energieversorgung. Berlin, 2010, p. 5 https://www.bundesregierung.de/ContentArchiv/DE/Archiv17/_Anlagen/2012/02/energiekonzept-final.pdf?__blob=publicationFile&v=5, access 8 April 2016

9. Relation of Technician Salary to Manager Salary at the Big Electricity Suppliers

Justification referring to Sustainability Concept	This indicator addresses the sustainability rule: 'Reduction of extreme income and wealth inequalities': While some people in Germany live in abundance, another part of the population lives below the poverty line. Wealth is also unevenly distributed around the world, especially between developing and industrialized countries. The uneven distribution of goods is at the root of many global and national problems [1]. Excessive wealth and income imbalances are to be avoided or ironed out because they are the cause of poverty and social marginalization.
Description	The information on the manager salary is taken from the official documents of the four biggest electricity suppliers, which are available on the Internet [2-25]. They include fees and basic salary plus vacation bonus and other payments and benefits as well as pensions and severance payments to members of the management board. The average income per person and year was set in relation to the average salary of the lowest income group in the energy sector. For a lack of measured data for the lowest income group, the information of the legal tariff register was used. Since this is not available for the whole of Germany, the data given for North Rhine-Westphalia is used in this example [26-27].
Unit	Technician salary in €/manager salary in €
Data	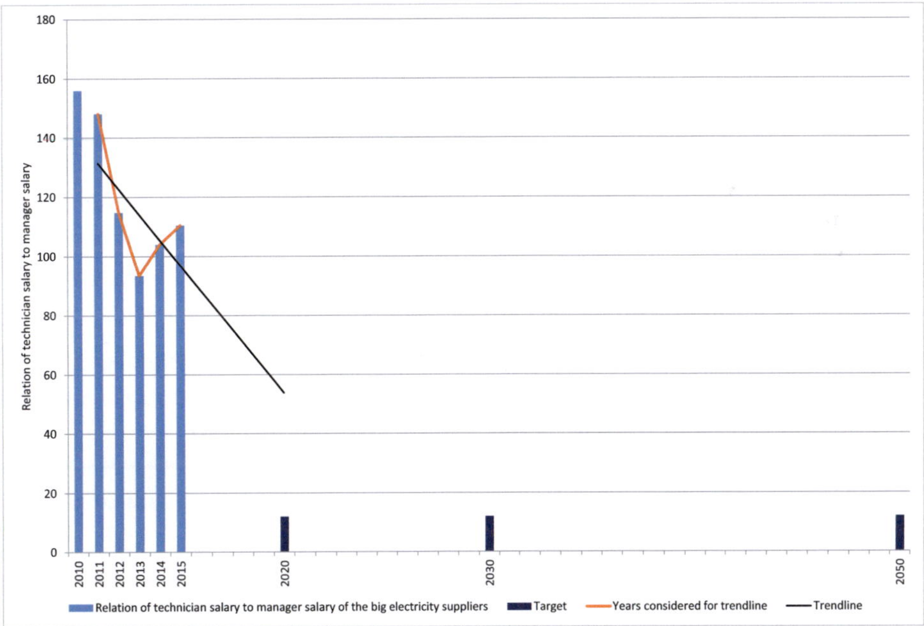 Figure 1: Relation of technician salary to manager salary at the four biggest electricity supplier (own diagram based on [2- 27])
Targets	2020: 1:12 2030: 1:12 2050: 1:12

Assessment	 Latest available value (2015): 1:110 [7, 13, 18, 25-27] The target is a reduction to a value of 12 % by 2020, which means a necessary reduction of 89 % compared to the reference value (the average value over the years 2011–2015). The extrapolated trend calculated on the reference value shows the expected change, which is a reduction of 53 % by 2020. The deviation is 41 %. Therfore, it is likely that the indicator will not reach the target for the year 2020 and gets a red traffic light. However, the indicator seems to be on track to reach the targets for 2030 and 2050. However, a closer look at the data shows that it is most likely that the current considerably decreasing trend will change to a smooth decreasing or stable trend when the 2011 value will be replaced by the 2016 value.
Comments Data	To determine the manager salary, it would be better to have the relevant data for the management boards and the technician of all electricity supply companies.
Comments Targets	For several years, there has been a discussion in Europe and the US about the different trends in the development of manager salaries and those of the other employees [29]. For example, the relation between the salaries of the members of the managing board and the average remuneration in a company has been in line in Germany for a long time and was at approximately 1:14 [30] in 1995. In the US, the relation between the incomes of ordinary workers to those of CEOs was 1:18 in 1965. In 2012, this relation had changed to 1:273 [31] and to 1:330 in 2015 [32]. Recently the discussion about the differences between managerial salaries and the incomes of employees was also intensified again in Germany [33]. A relation of 1:6.3 is considered fair in Germany according to a Harvard study [34]. In 2012, a citizens' initiative was launched in Switzerland, which demanded that the relation of worst paid employees to manager salaries should not be higher than 1:12. A referendum in Switzerland to make this a binding target at the national level was rejected in 2013 [28]. For the lack of an agreement on a binding target and against the background that a relation of this order has worked really well for decades until a few years ago, the Swiss target was adopted for the German electricity sector. In order to develop a better-justified target, it could be helpful to start a discussion – maybe not only in the electricity sector – about the publicly accepted relation between management salaries and salaries of the lowest income groups.
Literature	[1] Stieglitz, J. E.: Die wachsende Ungleichheit in unserer Gesellschaft. München. 2015 [2] Energie Baden-Württemberg – ENBW: Geschäftsbericht 2010. Karlsruhe. 2011 [3] Energie Baden-Württemberg – ENBW: Geschäftsbericht 2011. Karlsruhe. 2012 [4] Energie Baden-Württemberg – ENBW: Geschäftsbericht 2012. Karlsruhe. 2013 [5] Energie Baden-Württemberg – ENBW: Geschäftsbericht 2013. Karlsruhe. 2014 [6] Energie Baden-Württemberg – ENBW: Geschäftsbericht 2014. Karlsruhe. 2015 [7] Energie Baden-Württemberg – ENBW: Geschäftsbericht 2015. Karlsruhe. 2016 [8] E.ON: E.ON Geschäftsbericht 2010. Düsseldorf. 2011

[9] E.ON: E.ON Geschäftsbericht 2011. Düsseldorf. 2012

[10] E.ON: E.ON Geschäftsbericht 2012. Düsseldorf. 2013

[11] E.ON: E.ON Geschäftsbericht 2013. Düsseldorf. 2014

[12] E.ON: E.ON Geschäftsbericht 2014. Düsseldorf. 2015

[13] E.ON: E.ON Geschäftsbericht 2015. Düsseldorf. 2016

[14] RWE: Geschäftsbericht 2010. Essen. 2011

[15] RWE: Geschäftsbericht 2011. Essen. 2012

[16] RWE: Geschäftsbericht 2012. Essen. 2013

[17] RWE: Geschäftsbericht 2013. Essen. 2014

[18] RWE: Geschäftsbericht 2014. Essen. 2015

[19] RWE: Geschäftsbericht 2015. Essen. 2016

[20] VATTENFALL Europe AG: Das Jahr 2010 in Zahlen und Fakten. Berlin. 2011

[21] VATTENFALL Europe AG: Das Jahr 2011 in Zahlen und Fakten. Berlin. 2012

[22] VATTENFALL AB (publ): Geschäftsbericht 2012 inklusive Nachhaltigkeitsbericht. Stockholm. 2013

[23] VATTENFALL AB (publ): Geschäftsbericht 2013 inklusive Nachhaltigkeitsbericht. Stockholm. 2014

[24] VATTENFALL AB (publ): Geschäftsbericht 2014 inklusive Nachhaltigkeitsbericht. Stockholm. 2015

[25] VATTENFALL AB (publ): Geschäftsbericht 2015 inklusive Nachhaltigkeitsbericht. Stockholm. 2016

[26] Ministerium für Arbeit, Integration und Soziales des Landes NRW: Tarifinformationen 2013. Gas-, Wasser- und Elektrizitätsunternehmen, Tarifgruppe RWE. http://www.tarifregister.nrw.de/material/rwe.pdf, access 20 October 2013

[27] Ministerium für Arbeit, Integration und Soziales des Landes NRW: Tarifinformationen 2017. Gas-, Wasser- und Elektrizitätsunternehmen, Tarifgruppe RWE. http://www.tarifregister.nrw.de/material/rwe.pdf, access 10 March 2017

[28] Zumach, A.: Volksabstimmung in der Schweiz. Topverdiener werden nervös. In: taz, 23 May 2013, p. 9, http://www.taz.de/!5066916/, access 30 Nov 2016

[29] Hans-Böckler Stiftung: Manager to worker pay ratio. Mitbestimmungs-Report 25. Düsseldorf. 2016, https://www.boeckler.de/pdf/p_mbf_report_2016_25.pdf, access 15 March 2017

[30] Schwalbach, J.: Vergütungsstudie 2011. Vorstandsvergütung. Pay for Performance and Fair Pay in den DAX 30-Unternehmen. Berlin. 2011.

[31] Mishel, L.; Sabadish, N.: CEO Pay in 2012 was extraordinary high relative to typical workers and other high earners. Economic Policy Institute. Issue Brief 367. Washington DC. 2013, http://www.epi.org/publication/ceo-pay-2012-extraordinarily-high/, access 15 March 2017

	[32] Wirtschaftswoche: Konzernchefs verdienen gut 330 Mal mehr als ihre Arbeiter. 18. Mai 2016, http://www.wiwo.de/erfolg/management/gehaelter-konzernchefs-verdienen-gut-330-mal-mehr-als-ihre-arbeiter/13608668.html, access 15 March 2017 [33] Kraft, A.: Managergehälter: Kommt jetzt der Deckel per Gesetz? Magazin Mitbestimmung 03.03.2017, https://www.magazin-mitbestimmung.de/artikel/Manager-geh%C3%A4lter%3A+Kommt+jetzt+der+Deckel+per+Gesetz%3F@YIQrh-HowT2StAi4b5ei0Gg, access 10 March 2017 [34] Gavett, G.: CEOs Get Paid Too Much, According to Pretty Much Everyone in the World. Harvard Business Review. 23.09.2014, https://hbr.org/2014/09/ceos-get-paid-too-much-according-to-pretty-much-everyone-in-the-world, access 15 March 2017

10. Share of Renewable Energy in Gross Final Consumption of Energy

<table>
<tr><td>Justification referring to Sustainability Concept</td><td>This indicator addresses the sustainability rule: 'Sustainable use of renewable resources'. The usage rate of renewable resources should not exceed their own regeneration rate and should not compromise the performance and functionality of the relevant ecosystem. For example, biodiversity should be guaranteed through protecting habitats and nature as much as possible over the entire area.</td></tr>
<tr><td>Description</td><td>Calculation of the share of renewable energy in gross final energy consumption according to the definition of the EU directive 2009/28/EC on the promotion of the use of energy from renewable sources article 2 (f): 'gross final consumption of energy' means the energy commodities delivered for energy purposes to industry, transport, households, services including public services, agriculture, forestry and fisheries, including the consumption of electricity and heat by the energy branch for electricity and heat production and including losses of electricity and heat in distribution and transmission' [1].</td></tr>
<tr><td>Unit</td><td>%</td></tr>
<tr><td>Data</td><td>
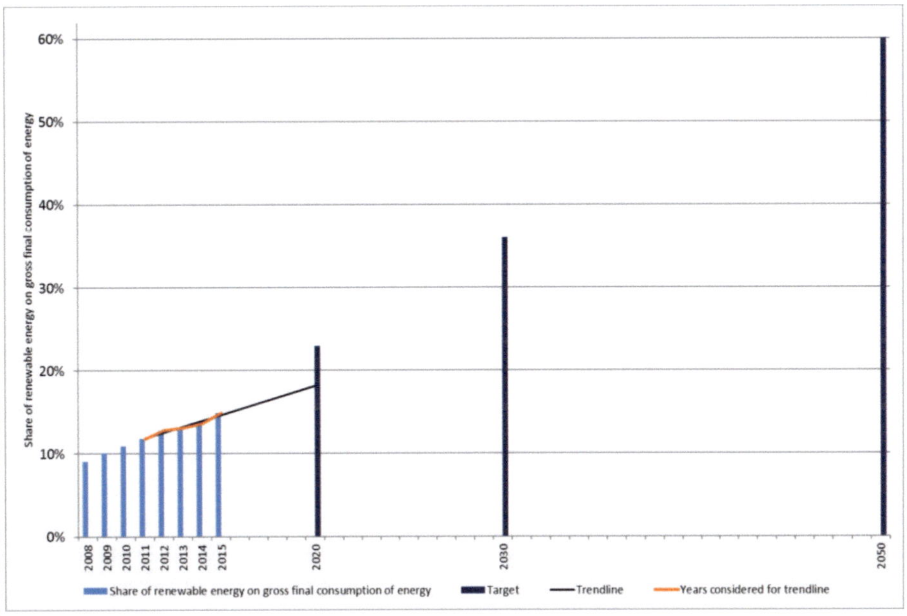

Figure 1: Share of renewable energy in gross final energy consumption (own diagram based on [2])</td></tr>
<tr><td>Targets</td><td>2020: 23 %
2030: 36 %
2050: 60 %</td></tr>
</table>

Assessment	 Latest available value (2015): 14.9 % [2] The target is an increase of 74 % by 2020 compared to the reference value over the previous 5 years from 2011 to 2015. The calculated trend results in an increase of only 14 % by 2020, which corresponds to a failure of 81 %. Therefore, a red traffic light is assigned since it is not likely that the target will be met in 2020.
Comments Data	The data refers to gross final consumption of energy. Data is also available for gross electricity consumption.
Comments Targets	The targets are based on the calculations made in [3]. They are slightly higher than those defined by the German Government in the energy concept 2010 [4] (18 % for 2020 and 30 % for 2030). For 2050, the values of the two sources are very similar (60 % in [3] and 61 % in [4]).
Literature	[1] EU: Directive 2009/28/EC of the European Parliament and of the council of 23 April 2009 on the promotion of the use of energy from renewable sources and amending and subsequently repealing Directives 2001/77/EC and 2003/30/EC. Brussels, 2009, http://eur-lex.europa.eu/legal-content/EN/ALL/?uri=CELEX%3A32009L0028, access 20 October 2016 [2] Bundesministerium für Wirtschaft und Energie: Fünfter Monitoring-Bericht zur Energiewende. Die Energie der Zukunft. November 2016, https://www.bmwi.de/Redaktion/DE/Publikationen/Energie/fuenfter-monitoring-bericht-energie-der-zukunft.html, access 22 February 2017 [3] DLR – Deutsches Zentrum für Luft- und Raumfahrt; IFNE – Ingenieurbüro für neue Energien; IWES – Fraunhofer Institut für Windenergie und Energiesystemtechnik: Langfristszenarien und Strategien für den Ausbau der erneuerbaren Energien in Deutschland bei Berücksichtigung der Entwicklung in Europa und global. Schlussbericht BMU, FKZ 03MAP146, March 2012, p. 144 http://www.dlr.de/dlr/Portaldata/1/Resources/bilder/portal/portal_2012_1/leitstudie2011_bf.pdf, access 23 March 2016 [4] Bundesregierung: Energiekonzept 2010. Berlin, 2010, p. 4 https://www.bundesregierung.de/ContentArchiv/DE/Archiv17/_Anlagen/2012/02/energiekonzept-final.pdf?__blob=publicationFile&v=5, access 23 March 2016

11. Area under Cultivation of Energy Crops

Justification referring to Sustainability Concept	This indicator addresses the sustainability rule: *'Sustainable use of renewable resources'*. The usage rate of renewable resources should not exceed their own regeneration rate and should not compromise the performance and functionality of the relevant ecosystem. For example, biodiversity should be guaranteed through protecting habitats and nature as much as possible over the entire area.
Description	For the cultivation of energy crops, part of the agricultural area must be used. This could lead to conflicts with other uses.
Unit	million ha
Data	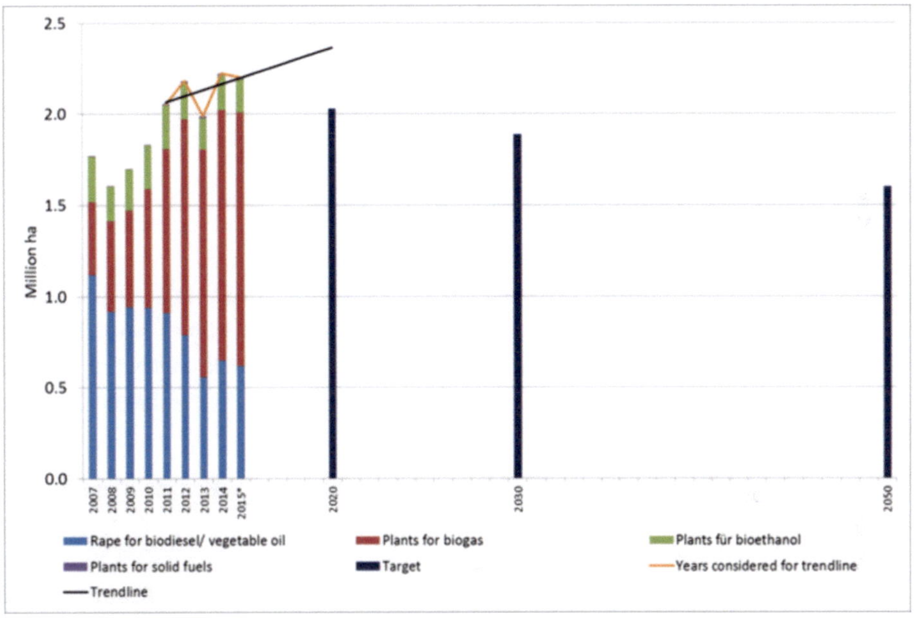 Figure 1: Area under cultivation of energy crops (own diagram based on [1]) *preliminary value
Targets	2020: 2.0 million ha 2030: 1.9 million ha 2050: 1.6 million ha
Assessment	 Latest available value (2015): 2.2 million ha [1] The extrapolated trend based on the previous 5 years (2011–2015) shows an increase in the area under cultivation of energy crops of about 11 % by 2020 compared to the reference value. The target for 2020, however, demands a decrease of 4.7 % compared to the average values measured for the years 2011 to 2015. Because the expected change goes in the wrong direction not reaching the target, a red traffic light is given.

<table>
<tr><td>Comments Data</td><td>Land is also used for conventional energy production (area used for plants, mining of brown coal), for RE plants (area used for wind energy plants, biogas plants), and for transmisison lines. Data for this kind of land use are given in [2]; however, the different kinds of land use cannot be summed up, because some of it can still be used for other purposes, some of it can be recultivated. Therefore, only land use for energy crops has been taken into consideration so far.

A land use conflict over agricultural land due to the competing goals of 'expansion of organic farming' and 'expansion of bioenergies' has already been documented in 2012 [3]. The competition between the utilization of agricultural land for food and feed, industrial crops and energy crops should also be kept in mind.

Importing biofuels and food will only shift land use conflicts to other countries. However, due to the fact that the focus of the analysis is on Germany, land use in other countries for the production of biomass that is exported to Germany has not been taken into consideration so far.</td></tr>
<tr><td>Comments Targets</td><td>According to [4], it is necessary to define limits for growing bioenergy crops or for terrestrial CO_2 storage in order to avoid land use conflicts. In this context, the following two points must be considered:

• The production of bioenergy sources and terrestrial CO_2 storage must not jeopardize the implementation of the WBGU area target of 10-20 % for nature conservation. As the current global total of protected areas is only 8.8 % (category I-VI areas), the conversion of natural ecosystems to land cultivated for bioenergy crops is rejected as a matter of principle.
• The production of food must take priority over the production of renewable energy sources or over CO_2 storage.

Based on these principles, it is possible to estimate the maximum area of crops grown for bioenergy that should be available globally or in specific regions. As a global benchmark, the WBGU recommends allocating no more than 3 % of the terrestrial area to such energy purposes. However, in order to avoid use conflicts with food and wood production and with the protection of natural ecosystems, detailed consideration of individual contingents is indispensable due to differences in local conditions.

As recommended in [4, p. 120], a maximum of 10 % of arable land and 10 % of pasture land should be available for the cultivation of energy crops in Europe. According to [4], the potential area for cultivation of energy crops for the European Area is 4.5 %. This value is used for calculating the potential area in contrast to the total land area (34.87 million ha [5]) in Germany. In doing so, this results in a maximum cultivation area for energy crops of about 1.6 million ha as target value for the year 2050.

From a starting point of 2.13 million ha (average over the years 2011 to 2015) and a target value of 1.6 million ha for the year 2050, the following targets are obtained for the years 2020 and 2030 by linear interpolation:

• 2020: 2.0 million ha (5.6 % of the land area of Germany)
• 2030: 1.9 million ha (5.4 % of the land area of Germany)
• 2050: 1.6 million ha (4.5 % of the land area of Germany)</td></tr>
</table>

Literature	[1] Statista: Anbaufläche von Energiepflanzen in Deutschland nach Art in den Jahren 2007 bis 2015 (in 1.000 Hektar). http://de.statista.com/statistik/daten/studie/153072/umfrage/anbauflaeche-von-energiepflanzen-in-deutschland-nach-sorten-seit-2007/, access 26 Oct 2016 [2] Löschel, A.; Erdmann, G; Staiß, F.; Ziesing, H..: Expertenkommission zum Monitoring-Prozess 'Energie der Zukunft' – Stellungnahme zum vierten Monitoring-Bericht der Bundesregierung für das Berichtsjahr 2014. Berlin, Münster, Stuttgart, November 2015, https://www.bmwi.de/Redaktion/DE/Downloads/M-O/monitoringbericht-energie-der-zukunft-stellungnahme-2014.pdf?__blob=publicationFile&v=3, access 03 Nov 2016 [3] Mayer, R.; Priefer, C.: Ökologischer Landbau und Bioenergieerzeugung. TAB-Arbeitsbericht Nr. 151, 2012 https://www.tab-beim-bundestag.de/de/pdf/publikationen/berichte/TAB-Arbeitsbericht-ab151.pdf, access 26 January 2016 [4] WBGU – Wissenschaftlicher Beirat der Bundesregierung Globale Umweltveränderungen: Welt im Wandel – Energiewende zur Nachhaltigkeit. March 2003 (p. 3) http://www.wbgu.de/fileadmin/user_upload/wbgu.de/templates/dateien/veroeffentlichungen/hauptgutachten/jg2003/wbgu_jg2003.pdf, access 12 February 2015 [5] Lexas. Flächendaten aller Staaten der Erde. http://www.laenderdaten.de/geographie/flaeche_staaten.aspx, access 30 Nov 2016

12. Unused Renewable Electricity due to Management Measures

Justification referring to Sustainability Concept	This indicator addresses the sustainability rule: 'Sustainable use of renewable resources'. The usage rate of renewable resources should not exceed their own regeneration rate and should not compromise the performance and functionality of the relevant ecosystem. For example, biodiversity should be guaranteed through protecting habitats and nature as much as possible over the entire area.
Description	To avoid or eliminate bottlenecks within the electricity grid, it is sometimes necessary to force power plants to reduce the feed-in of electricity. For renewable energy plants, this means that some amount of renewable energy that would normally be used, cannot be used. This reduces the amount of renewable electricity produced.
Unit	GWh
Data	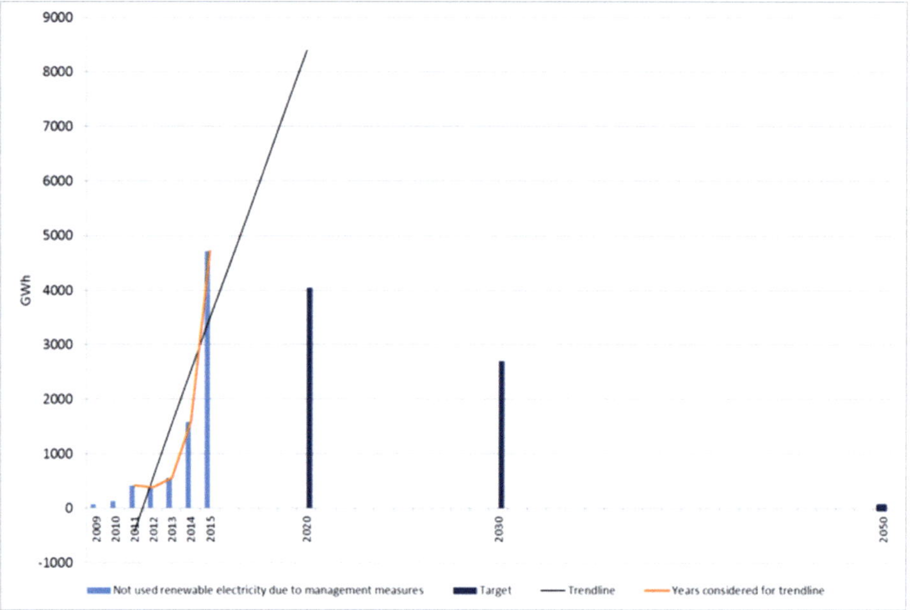 Figure 1: Renewable electricity not used due to management measures in GWh (own diagram based on [1-3])
Targets	2020: 4,047 GWh 2030: 2,698 GWh 2050: 0 GWh
Assessment	 Latest available value (2015): 4,722 GWh [1] The trend shows a strong upward tendency, which is mainly due to the sharp increase in the years 2014 and 2015. However, the target is a significant reduction in the loss of electricity production. A red traffic light is assigned because the achievement of the target of 2020 is not likely.

Comments Data	As long as the share of inflexible nuclear and coal-fired power plants in the grid is high and at the same time the expansion of renewable energies continues, the undesired shut-down of renewable energy systems will occur again and again. One way to minimize the amount of 'electricity not used' is to reduce the number of sluggish power plants significantly and to convert the fossil part of the power plants to flexible gas-fired power plants. Other ways will be the better connection of grids, the inclusion of flexible consumers and the integration of storage capacities, including Power-to-X technologies.
Comments Targets	The target is to reduce the amount of unused renewable generated electricity to '0' by 2050. This target can be achieved by connecting different subnetworks of the power grid, integrating controllable consumers into the grid, and by increasing the availbale power storage capacities. These measures will ensure that high amounts of electricity produced by renewables accruing in times of low consumption can increasingly be used directly or after storage.
Literature	[1] Bundesnetzagentur, Bundeskartellamt: Monitoringbericht 2016. Bonn, p. 5. https://www.bundesnetzagentur.de/DE/Sachgebiete/ElektrizitaetundGas/Unternehmen_Institutionen/DatenaustauschundMonitoring/Monitoring/Monitoringberichte/Monitoring_Berichte_node.html, access 25 January 2017 [2] Bundesnetzagentur, Bundeskartellamt: Monitoringbericht 2015. Bonn, p. 5. https://www.bundesnetzagentur.de/DE/Sachgebiete/ElektrizitaetundGas/Unternehmen_Institutionen/DatenaustauschundMonitoring/Monitoring/Monitoringberichte/Monitoring_Berichte_node.html, access 20 October 2016 [3] Bundesnetzagentur, Bundeskartellamt: Monitoringbericht 2014. Bonn, p. 81. https://www.bundesnetzagentur.de/DE/Sachgebiete/ElektrizitaetundGas/Unternehmen_Institutionen/DatenaustauschundMonitoring/Monitoring/Monitoringberichte/Monitoring_Berichte_node.html, access 20 October 2016

13. Use of Primary Energy

<table>
<tr><td>Justification referring to Sustainability Concept</td><td>This indicator addresses the sustainability rule 'Sustainable use of non-renewable resources': The range of proven non-renewable resources must be maintained over time. Related to the energy system, this includes the range of fossil and nuclear energy carriers as well as raw materials for building energy infrastructures. Improving energy efficiency and productivity and reduced consumption volumes can contribute to extending the ranges.</td></tr>
<tr><td>Description</td><td>Amount of primary energy used within one year.</td></tr>
<tr><td>Unit</td><td>PJ/a</td></tr>
<tr><td>Data</td><td>
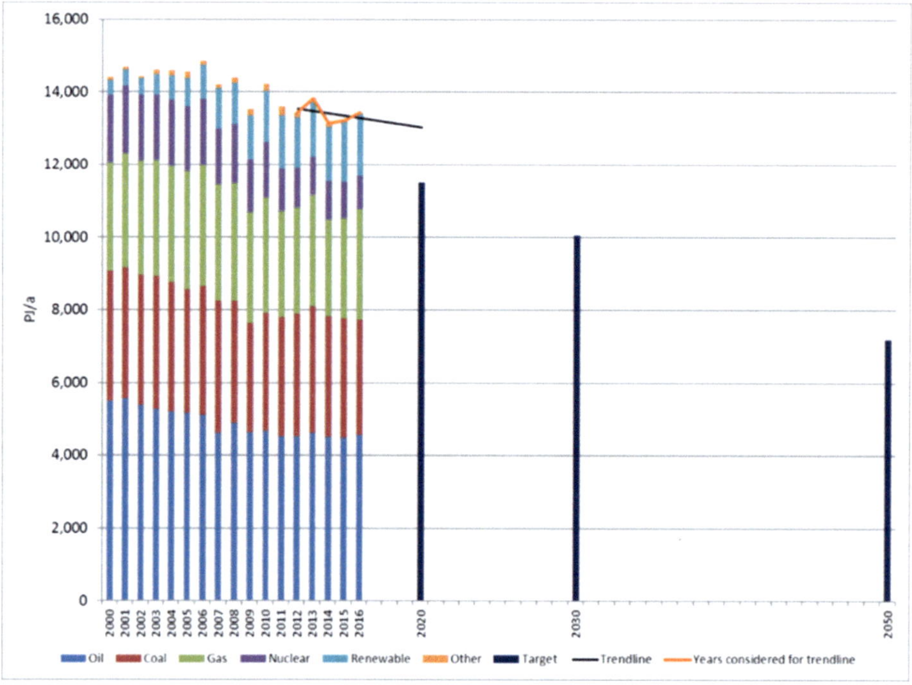

Figure 1: Amount of primary energy used within one year in PJ/a (own diagram based on [1-2])</td></tr>
<tr><td>Targets</td><td>2020: 11,504 PJ/a
2030: 10,066 PJ/a
2050: 7,190 PJ/a</td></tr>
<tr><td>Assessment</td><td>

Value for the reference year (2008): 14,380 PJ/a [1]

Latest available value (2016): 13,427 PJ/a [2]

The target is a reduction of 14.2 % by 2020 compared to the reference value (the average value over the previous 5 years 2012 to 2016). The calculated trend results in an expected reduction of only 2.9 % by 2020, which corresponds to a deviation to the necessary change of 79.7 %.

This results in a red traffic as the target achievement in 2020 is not likely.</td></tr>
</table>

Comments Data	The values in Figure 1 are not temperature-corrected.
Comments Targets	In [3] it is defined that the use of primary energy has to be reduced by 20 % by 2020 and by 50 % by 2050 compared to the value in 2008. Based on these rules, the targets are calculated. The value for 2030 is calculated via interpolation between 2020 and 2050.
Literature	[1] AG Energiebilanzen: Auswertungstabellen zur Energiebilanz für die Bundesrepublik Deutschland 1990 bis 2015 [2] http://www.ag-energiebilanzen.de/10-0-Auswertungstabellen.html, access 14 February 2017 [3] AG Energiebilanzen: Pressedienst Nr. 05/2016 – Energieverbrauch legt 2016 zu. http://www.ag-energiebilanzen.de/index.php?article_id=22&archiv=18&year=2016, access 14 February 2017 [4] BMWi – Bundesministerium für Wirtschaft und Technologie; BMU – Bundesministerium für Umwelt, Naturschutz und Reaktorsicherheit: Energiekonzept für eine umweltschonende, zuverlässige und bezahlbare Energieversorgung. Berlin, 2010, p. 5 http://www.e2a.de/data/files/energiekonzept_bundesregierung.pdf, access 8 April 2016

14. Specific Final Energy Consumption of Households for Heating (Temperature-corrected)

<table>
<tr><td>Justification referring to Sustainability Concept</td><td>This indicator addresses the sustainability rule 'Sustainable use of non-renewable resources': The range of proven non-renewable resources must be maintained over time. Related to the energy system, this includes the range of fossil and nuclear energy carriers as well as raw materials for building energy infrastructures. Improving energy efficiency and productivity and reduced consumption volumes can contribute to extending the ranges.</td></tr>
<tr><td>Description</td><td>Temperature-corrected specific final energy consumption of households (MJ per m^2 living space) is a measure for the efficiency of the heating behavior of buildings.</td></tr>
<tr><td>Unit</td><td>MJ/m^2 living space</td></tr>
<tr><td>Data</td><td>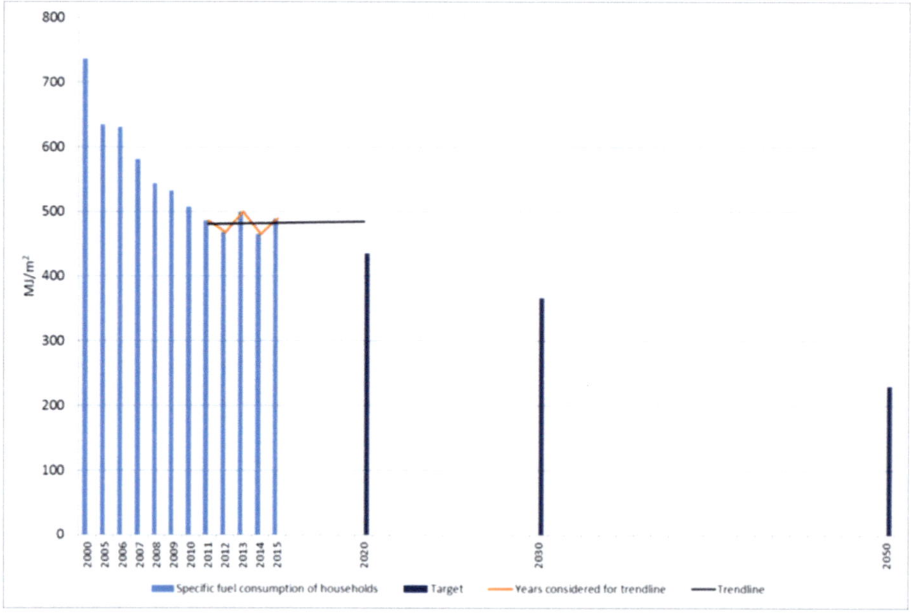

Figure 1: Temperature-corrected specific energy consumption of households for heating (own diagram based on [1])
(Note: 500 MJ equals approx. 139 kWh)</td></tr>
<tr><td>Targets</td><td>2020: 435 MJ/m^2
2030: 367 MJ/m^2
2050: 230 MJ/m^2</td></tr>
<tr><td>Assessment</td><td>
Value for the reference year (2008): 544 MJ/m^2 [1]
Latest available value (2015): 490 MJ/m^2 [1]
The extrapolated trend based on the previous 5 years (2011–2015) shows an increase in the specific energy consumption of households for heating of 0.5 % by 2020 compared to the reference value.</td></tr>
</table>

	The necessary change to achieve the target requires a reduction of 9.7 % compared to the reference value. This results in a deviation of 105 % and a red traffic light.
Comments Data	Data are taken from [1] and are temperature-corrected, that means that data do not include the influence of warm and cold years, respectively. Data given in [1] are based on the total energy consumption of households divided by the total living space for the given year.
Comments Targets	According to the German Federal Government, energy consumption for heating of buildings should be reduced by 20 % by 2020 compared to 2008. Assuming a constant living space, the target value for the year 2020 is 435 MJ/m^2. In [2], a value of 230 PJ/m^2 is suggested for the year 2050, corresponding to a reduction of 60 % compared to 2008. The value for 2030 results from linear interpolation between the values for 2020 and 2050. As a result, the following target values have been defined: • 2020: 435 PJ/m^2 (20 % reduction compared to 2008) • 2030: 367 PJ/m^2 (33 % reduction compared to 2008) • 2050: 230 PJ/m^2 (60 % reduction compared to 2008)
Literature	[1] BMWi – Bundesministerium für Wirtschaft und Energie: Die Energie der Zukunft. Fünfter Monitoring-Bericht zur Energiewende. Berichtsjahr 2015, Dezember 2016 https://www.bmwi.de/Redaktion/DE/Publikationen/Energie/fuenfter-monitoring-bericht-energie-der-zukunft.pdf?__blob=publicationFile&v=23, access 2 February 2017 [2] DLR – Deutsches Zentrum für Luft- und Raumfahrt; IfNE – Ingenieurbüro für neue Energien; IWES – Fraunhofer Institut für Windenergie und Energiesystemtechnik: Langfristszenarien und Strategien für den Ausbau der erneuerbaren Energien in Deutschland bei Berücksichtigung der Entwicklung in Europa und global. Schlussbericht BMU – FKZ 03MAP146, March 2012 http://www.dlr.de/dlr/Portaldata/1/Resources/bilder/portal/portal_2012_1/leitstudie2011_bf.pdf, access 4 July 2016

15. Final Energy Consumption in the Transport Sector

<table>
<tr><td>Justification referring to Sustainability Concept</td><td>This indicator addresses the sustainability rule 'Sustainable use of non-renewable resources': The range of proven non-renewable resources must be maintained over time. Related to the energy system, this includes the range of fossil and nuclear energy carriers as well as raw materials for building energy infrastructures. Improving energy efficiency and productivity and reduced consumption volumes can contribute to extending the ranges.</td></tr>
<tr><td>Description</td><td>Final energy consumption in the transport sector includes the consumption of gasoline, diesel, kerosene, electricity (but not electricity for road transport) as well as fuels from biomass for rail, road, air and ship transport (coastal shipping and inland water transport).</td></tr>
<tr><td>Unit</td><td>PJ</td></tr>
<tr><td>Data</td><td>
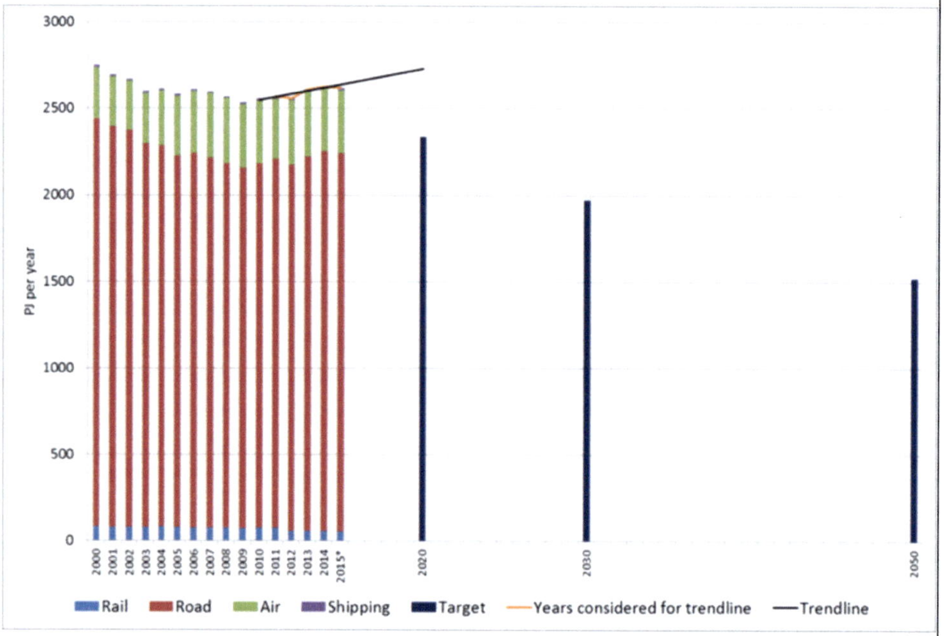

Figure 1: Final energy consumption in the transport sector (own diagram based on [1])

* preliminary values
</td></tr>
<tr><td>Targets</td><td>2020: 2,337 PJ
2030: 1,973 PJ
2050: 1,521 PJ</td></tr>
<tr><td>Assessment</td><td>

Value for the reference year (2005): 2,586 PJ [1]

Latest available value (2015): 2,619 PJ [1]

The trend based on the previous 5 years (2011–2015) shows an increase in final energy consumption in the transport sector of 4.6 % by 2020 compared to the reference value (the average value over the years 2011 to 2015). However, the target for 2020 requires a reduction of 10 % compared to the reference value. The deviation is 146 % and a red traffic light is given.
</td></tr>
</table>

Comments Data	Data are taken from [1] and [2] taking into account the transport sectors of rail, road, air, and shipping.
Comments Targets	Final energy consumption in the transport sector should be reduced by 10 % by 2020, compared to 2005 [2]. This would result in a final energy consumption in the transport sector of 2,327 PJ in 2020. In [3], the following data are given based on the characteristics of the transport sector in the years 2020, 2030 and 2050 (vehicle fleet, km driven, etc.): • 2020: 2,337 PJ (a reduction of 10 % compared to 2005) • 2030: 1,973 PJ (a reduction of 24 % compared to 2005) • 2050: 1,521 PJ (a reduction of 41 % compared to 2005)
Literature	[1] BMWi – Bundesministerium für Wirtschaft und Energie: Datenübersicht zum Fünften Monitoringbericht. Fünfter Monitoringbericht. Die Energie der Zukunft. November 2016, https://www.bmwi.de/Redaktion/DE/Publikationen/Energie/fuenfter-monitoring-bericht-energie-der-zukunft.html, access 22 February 2017 [2] BMWi – Bundesministerium für Wirtschaft und Energie: Datenübersicht Fortschrittsbericht 2014. Ein gutes Stück Arbeit – Die Energie der Zukunft. Vierter Monitoringbericht zur Energiewende. November 2015, https://www.bmwi.de/Redaktion/DE/Publikationen/Energie/vierter-monitoring-bericht-energie-der-zukunft.pdf?__blob=publicationFile&v=22 access 13 June 2016 [3] DLR – Deutsches Zentrum für Luft- und Raumfahrt; IfNE – Ingenieurbüro für neue Energien; IWES – Fraunhofer Institut für Windenergie und Energiesystemtechnik: Langfristszenarien und Strategien für den Ausbau der erneuerbaren Energien in Deutschland bei Berücksichtigung der Entwicklung in Europa und global. Schlussbericht BMU – FKZ 03MAP146, March 2012 http://www.dlr.de/dlr/Portaldata/1/Resources/bilder/portal/portal_2012_1/leitstudie2011_bf.pdf, access 13 June 2016

16. Modal Split in the Transport Sector

Justification referring to Sustainability Concept	This indicator addresses the sustainability rule *'Sustainable use of non-renewable resources'*: The range of proven non-renewable resources must be maintained over time. Related to the energy system, this includes the range of fossil and nuclear energy carriers as well as raw materials for building energy infrastructures. Improving energy efficiency and productivity and reduced consumption volumes can contribute to extending the ranges.
Description	The modal split is applied to the following means of transport: individual transport, public transport on the road or on rail tracks, transport by air, bicycles, walking. Transport by bicycle or walking is considered as non-motorized. Transport by tram, train and bus is considered as public transport. Transport performance is measured in person-km.
Unit	%
Data	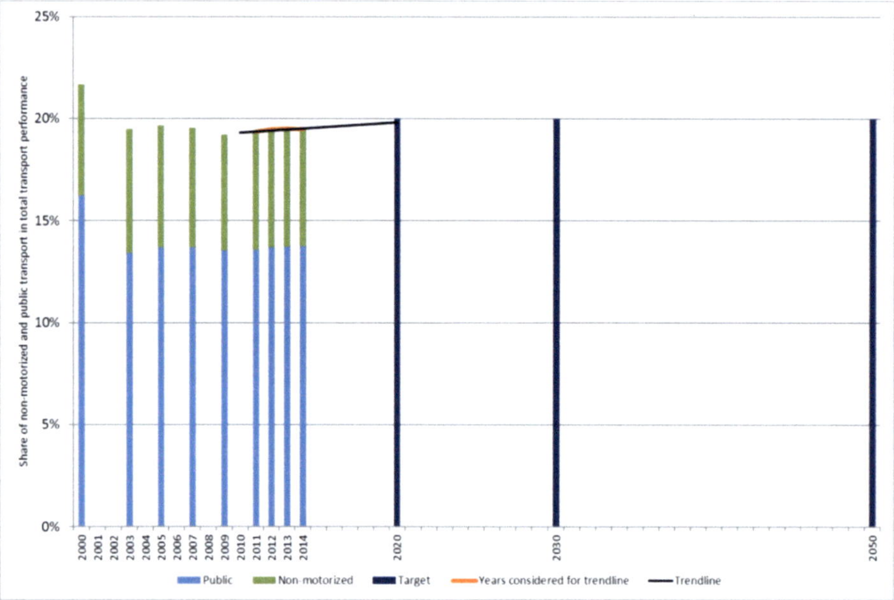 Figure 1: Share of non-motorized and public transport in total transport performance (own diagram based on [1])
Targets	2020: 20 % 2030: 20 % 2050: 20 %
Assessment	 Latest available value (2014): 19.5 % [1] There is nearly no change in the modal split since 2003. However, the calculated trend based on data for the previous five years (2009 to 2014) shows a slight increase. The increase of the modal split until 2020 should be 2.8 % (compared to the average value over the years 2009 to 2014), whereas the trend only results in an increase by 1.8 %, which is a deviation of about 35 % and a yellow traffic light.

Comments Data	No data are available for 2010. Therefore, data for 2010 have been calculated as an average betweeen 2009 and 2011. The modal split is applied to the transport volume ('Verkehrsaufwand') of the means of transport outlined above. The transport volume for individuals represents the product of the number of persons multiplied with the transport distance. The transport volume of transported goods is not taken into consideration.
Comments Targets	Since 1990, the modal split has changed little. An analysis of the data on transport services in [2] until 2050 shows that also for this period no change in the modal split is expected. The necessary reduction of energy consumption in the transport sector by 2050 is, by contrast, achieved by significant gains in efficiency in this sector. Therefore, the current modal split of 20 % is maintained as target value for the years 2020, 2030 and 2050.
Literature	[1] BMVI – Bundesministerium für Verkehr und digitale Infrastruktur: Verkehr in Zahlen 2016/2017. September 2016 http://www.bmvi.de/SharedDocs/DE/Anlage/VerkehrUndMobilitaet/verkehr-in-zahlen-pdf-2016-2017.pdf?__blob=publicationFile, access 15 Dec. 2016 [2] DLR – Deutsches Zentrum für Luft- und Raumfahrt; IfNE – Ingenieurbüro für neue Energien; IWES – Fraunhofer Institut für Windenergie und Energiesystemtechnik: Langfristszenarien und Strategien für den Ausbau der erneuerbaren Energien in Deutschland bei Berücksichtigung der Entwicklung in Europa und global. Schlussbericht BMU – FKZ 03MAP146, March 2012 http://www.dlr.de/dlr/Portaldata/1/Resources/bilder/portal/portal_2012_1/leitstudie2011_bf.pdf, access 21 June 2016

17. Number of Electric Vehicles

Justification referring to Sustainability Concept	This indicator addresses the sustainability rule *'Sustainable use of non-renewable resources'*: The range of proven non-renewable resources must be maintained over time. Related to the energy system, this includes the range of fossil and nuclear energy carriers as well as raw materials for building energy infrastructures. Improving energy efficiency and productivity and reduced consumption volumes can contribute to extending the ranges.
Description	Vehicles with electric drive that obtain part or all of their energy from the electric grid are more energy efficient than vehicles with combustion engine. In addition, direct emissions of CO_2, acidogenic gases and particles are reduced or there are no direct emissions of CO_2, acidogenic gases and particles from vehicles with electric drive only. However, life cycle assessments have shown that, with existing technologies, electrically driven cars under special circumstances have a lower climate impact when they use electricity from renewable or nuclear sources. Taking into account the whole supply chain, electric cars of today in some cases achieve worse results for indicators such as human toxicity potential than conventional cars [1]. Most of these effects will be reduced in the future, for example, when batteries are reused or recycled, other materials for batteries are found or a higher percentage of the power is generated by renewables.
Unit	Number of multi-track vehicles with plug-in electric drive
Data	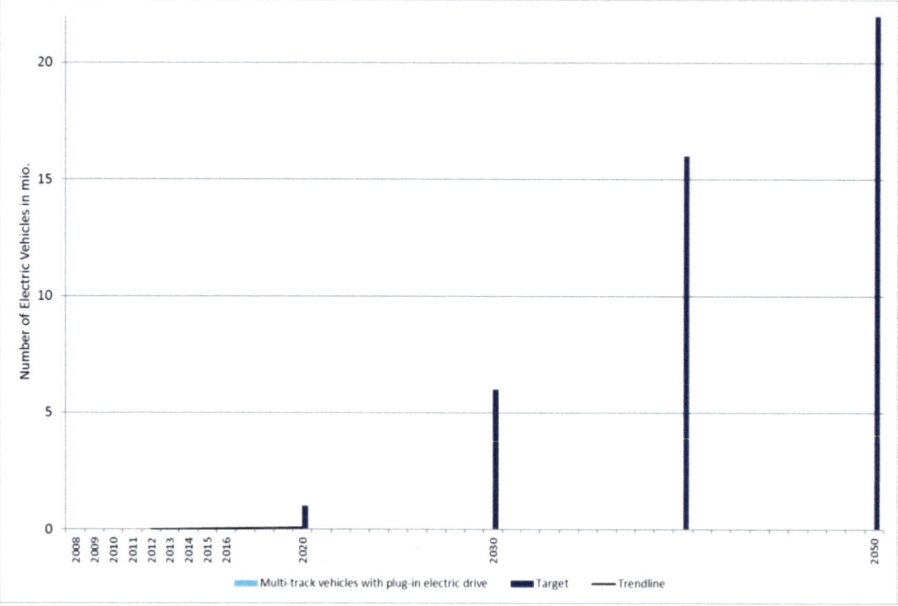 Figure 1: Number of multi-track vehicles with electric drive and plug-in hybrid (own diagram based on [2])
Targets	2020: 1,000,000 multi-track vehicles with plug-in electric drive [3] 2030: 6,000,000 multi-track vehicles with plug-in electric drive [3] 2050: 22,000,000 multi-track vehicles with plug-in electric drive [4]

Assessment	 Latest available value (2016): 41,460 [2] The calculated trend based on the previous 5 years (2012–2016) indicates a number of approx. 55,071 electric vehicles in 2020. This means that in 2020 only 5.5 % of the target value would be realized. A red traffic light is given, as it is likely that the target for 2020 will be missed.
Comments Data	Multi-track electric vehicles include cars, buses, trucks, tractors, and other motor vehicles (fire services etc.). The number of new electric passenger cars (for other vehicles no numbers for this years are available) registered in Germany from 2000 to 2008 ranged between 8 and 109 per year. From 2009 on, the number of electric driven multi-track vehicles has steadily increased [5]: from 162 (in 2009), 541 (in 2010), 2,154 (in 2011), 2,956 (in 2012), 6,051 (in 2013), 8,522 (in 2014) to 12,363 (in 2015).
Comments Targets	The Federal Government's energy concept sets the targets of 1,000,000 electric vehicles in 2020 and 6,000,000 in 2030 [3]. In [4] a number of 22,000,000 vehicles in 2050 is set as target for future development of electric vehicles.
Literature	[1] Bauer, C.; Hofer, J.; Althaus, H.-J.; Del Duce, A.; Simons, A.: The environmental performance of current and future passenger vehicles: Life Cycle Assessment based on a novel scenario analysis framework. Applied Energy 2015, 157(C), pp. 871–883 [2] BMWi – Bundesministerium für Wirtschaft und Energie: Die Energie der Zukunft. Datenübersicht zum Fünften Monitoring-Bericht zur Energiewende, Berichtsjahr 2015 https://www.bmwi.de/Redaktion/DE/Artikel/Energie/monitoring-prozess.html, access 1 February 2017 [3] BMWi - Bundesministerium für Wirtschaft und Energie; BMU - Bundesministerium für Umwelt, Naturschutz und Reaktorsicherheit : Energiekonzept für eine umweltschonende, zuverlässige und bezahlbare Energieversorgung, Berlin, 2010 https://www.bmwi.de/BMWi/Redaktion/PDF/E/energiekonzept-2010,property=pdf,bereich=bmwi2012,sprache=de,rwb=true.pdf, access 22 June 2016 [4] DLR – Deutsches Zentrum für Luft- und Raumfahrt; IfNE – Ingenieurbüro für neue Energien; IWES – Fraunhofer Institut für Windenergie und Energiesystemtechnik: Langfristszenarien und Strategien für den Ausbau der erneuerbaren Energien in Deutschland bei Berücksichtigung der Entwicklung in Europa und global. Schlussbericht BMU – FKZ 03MAP146, March 2012, p. 131 www.dlr.de/dlr/Portaldata/1/Resources/bilder/portal/portal_2012_1/leitstudie2011_bf.pdf, access 23 June 2016 [5] Kraftfahrt-Bundesamt: Fahrzeugzulassungen (FZ), Neuzulassungen von Kraftfahrzeugen nach Umwelt-Merkmalen Jahr 2015. FZ 14 Flensburg, 2016 http://www.kba.de/SharedDocs/Publikationen/DE/Statistik/Fahrzeuge/FZ/2015/fz14_2015_pdf.pdf?__blob=publicationFile&v=3, access 1 February 2017

18. Final Energy Productivity of the German Economy

Justification referring to Sustainability Concept	This indicator addresses the sustainability rule *'Sustainable use of non-renewable resources'*: The range of proven non-renewable resources must be maintained over time. Related to the energy system, this includes the range of fossil and nuclear energy carriers as well as raw materials for building energy infrastructures. Improving energy efficiency and productivity and reduced consumption volumes can contribute to extending the ranges.
Description	Energy productivity describes how much value (GDP, Gross Domestic Products in €) is produced in Germany per GJ of energy used
Unit	€ per GJ
Data	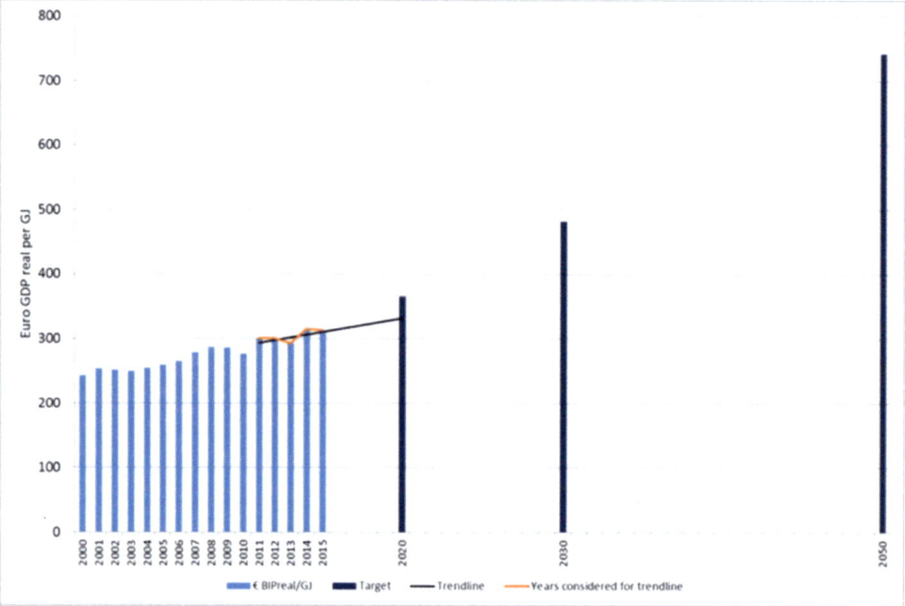 Figure 1: Final energy productivity of the German economy (real GDP, temperature- and stock-adjusted) (own diagram based on [1])
Targets	2020: 366 €/GJ 2030: 482 €/GJ 2050: 743 €/GJ
Assessment	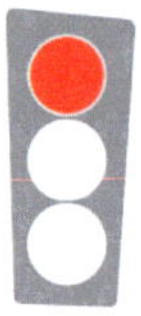 Value for the reference year (2010): 277 € GDP real per GJ [1] Latest available value (2015): 314 € GDP real per GJ [1] The calculated trend based on the previous 5 years (2011–2015) shows an increase of the final energy productivity of 9 % by 2020 compared to the reference value (the average value over the years 2011 to 2015). The target for 2020 requires an increase of 20 % compared to reference value. This results in a deviation of approx. 56 % in the year 2020 and a red traffic light.

Comments Data	The data take into account the final energy consumption and real GDP. These values are temperature- and stock-adjusted. Data are also available for primary energy consumption and for primary energy productivity.
Comments Targets	Objective of the Federal Government for the period until 2050 is an improvement of 2.1 %/a. Starting in 2011, an increase in productivity of 2.1 %/a would lead to an increase by the factor 1.23 by 2020, by the factor 1.52 by 2030, and by the factor 2.3 by 2050 (each based on the value of 2010). Even though, according to [2], the increase in (primary) energy productivity of 1.65 %/a in the long-term average (1990–2010) was higher than the average growth rate of the GDP of 1.35 %/a, this minor difference is not sufficient for substantial reductions in consumption. Therefore [2] assumes an increase of the average growth rate of (primary) energy productivity to an average of 2.8 %/a between 2011 and 2030 and to an average of 2.2 %/a between 2031 and 2050. Starting in 2011, an increase in productivity of 2.8 %/a for the period 2011 to 2030 and of 2.2 %/a for the period 2031 to 2050 would lead to an increase by the factor 1.32 by 2020 and by the factor 2.68 by 2050 (each based on the value of 2010). These values, which are more ambitious than those of the Federal Government, are also used for the target values. This also guarantees the consistency with the target values of other indicators which are also taken from [2]. However, it must be noted that the growth rates given in [2] refer to primary energy productivity while the indicator refers to final energy productivity. Therefore the chosen target values might be more ambitious, since, for example, an increase of the share of solar energy in total energy production with otherwise identical boundary conditions leads to an increase in primary energy productivity, but not to an increase in final energy productivity. This results in the following target values (initial value 2010: 277 € per GJ) • 2020: 366 €/GJ (increase by the factor 1.32 compared to 2010) • 2030: 482 €/GJ (increase by the factor 1.74 compared to 2010) • 2050: 743 €/GJ (increase by the factor 2.68 compared to 2010)
Literature	[1] BMWi – Bundesministerium für Wirtschaft und Energie: Die Energie der Zukunft. Fünfter Monitoring-Bricht zur Energiewende. Berichtsjahr 2015, 2016 https://www.bmwi.de/Redaktion/DE/Publikationen/Energie/fuenfter-monitoring-bericht-energie-der-zukunft.pdf?__blob=publicationFile&v=23, access: 1 February 2017 [2] DLR – Deutsches Zentrum für Luft- und Raumfahrt; IfNE – Ingenieurbüro für neue Energien; IWES – Fraunhofer Institut für Windenergie und Energiesystemtechnik: Langfristszenarien und Strategien für den Ausbau der erneuerbaren Energien in Deutschland bei Berücksichtigung der Entwicklung in Europa und global. Schlussbericht BMU, FKZ 03MAP146, March 2012 www.dlr.de/dlr/Portaldata/1/Resources/bilder/portal/portal_2012_1/leitstudie2011_bf.pdf, access 29 June 2016

19. Final Energy Productivity of the Industry

Justification referring to Sustainability Concept	This indicator addresses the sustainability rule *'Sustainable use of non-renewable resources'*: The range of proven non-renewable resources must be maintained over time. Related to the energy system, this includes the range of fossil and nuclear energy carriers as well as raw materials for building energy infrastructures. Improving energy efficiency and productivity and reduced consumption volumes can contribute to extending the ranges.
Description	Energy productivity of the German industry describes how much Gross Value Added (GVA) is produced in the industry per GJ of energy.
Unit	€ GVA per GJ
Data	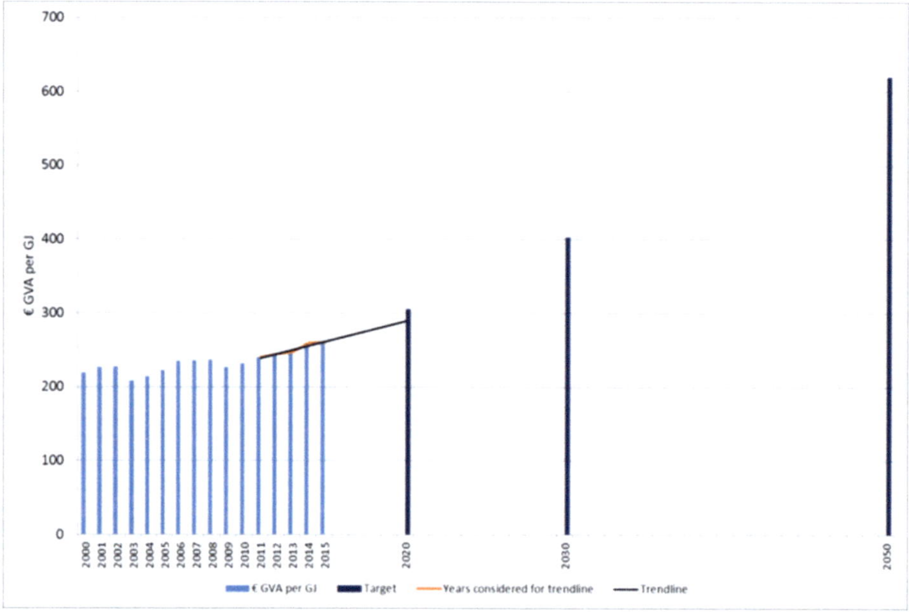 Figure 1: Final energy productivity of the German industry (own diagram based on [1])
Targets	2020: 306 €/GJ 2030: 403 €/GJ 2050: 621 €/GJ
Assessment	 Value for the reference year (2010): 232 € GVA per GJ [1] Latest available value (2015): 261 € GVA per GJ [1] The trend calculated based on the previous 5 years (2011–2015) shows an increase of the final energy productivity of 16 % by 2020 compared to the reference value (the average value over the years 2011 to 2015). The target value for 2020 requires an increase of 22 % compared to reference value. The deviation of 23 % results in a yellow traffic light.
Comments Data	The data take into account final energy consumption and real GVA. Data are also available for primary energy consumption and thus for primary energy productivity.

Comments Targets	Objective of the Federal Government for the period until 2050 is an improvement of the overall energy productivity of 2.1 %/a [2]. Starting in 2011, an increase in productivity of 2.1 %/a would lead to an increase by the factor 1.23 by 2020, by the factor 1.52 by 2030, and by the factor 2.3 by 2050 (each based on the value of 2010). Even though, according to [3], the increase in (primary) energy productivity of 1.65 %/a in the long-term average (1990–2010) was higher than the average growth rate of the GDP of 1.35 %/a, this minor difference is not sufficient for substantial reductions in consumption. Therefore [3] assumes an increase of the average growth rate of (primary) energy productivity to an average of 2.8 %/a between 2011 and 2030 and to an average of 2.2 %/a between 2031 and 2050. Starting in 2011, an increase in productivity of 2.8 %/a for the period 2011 to 2030 and of 2.2 % for the period 2031 to 2050 would lead to an increase by the factor 1.32 by 2030 and by the factor 2.68 by 2050 (each based on the value of 2010). These values, which are more ambitious than those of the Federal Government, are also used for the industry's target values. This also guarantees the consistency with the target values of other indicators, which are also taken from [3]. However, it must be noted that the growth rates given in [3] refer to primary energy productivity while the indicator refers to final energy productivity. Therefore, the chosen target values might be more ambitious, since, for example, an increase of the share of solar energy in total energy production with otherwise identical boundary conditions leads to an increase in primary energy productivity, but not to an increase in final energy productivity.
Literature	[1] BMWi – Bundesministerium für Wirtschaft und Energie: Gesamtausgabe der Energiedaten – Datensammlung des BMWi. Zahlen und Fakten Energiedaten. 2018. www.bmwi.de/Redaktion/DE/Binaer/Energiedaten/energiedaten-gesamt-xls.xls?__blob=publicationFile&v=73, access 7 March 2018 [2] BMWi – Bundesministerium für Wirtschaft und Energie: Die Energie der Zukunft. Fünfter Monitoring-Bricht zur Energiewende. Berichtsjahr 2015, 2016 https://www.bmwi.de/Redaktion/DE/Publikationen/Energie/fuenfter-monitoring-bericht-energie-der-zukunft.pdf?__blob=publicationFile&v=23, access 1 February 2017 [3] DLR – Deutsches Zentrum für Luft- und Raumfahrt; IfNE – Ingenieurbüro für neue Energien; IWES – Fraunhofer Institut für Windenergie und Energiesystemtechnik: Langfristszenarien und Strategien für den Ausbau der erneuerbaren Energien in Deutschland bei Berücksichtigung der Entwicklung in Europa und global. Schlussbericht BMU, FKZ 03MAP146, March 2012 http://www.dlr.de/dlr/Portaldata/1/Resources/bilder/portal/portal_2012_1/leitstudie2011_bf.pdf, access 29 June 2016

20. Final Energy Productivity of Trade, Commerce and Services

Justification referring to Sustainability Concept

This indicator addresses the sustainability rule *'Sustainable use of non-renewable resources'*: The range of proven non-renewable resources must be maintained over time. Related to the energy system, this includes the range of fossil and nuclear energy carriers as well as raw materials for building energy infrastructures. Improving energy efficiency and productivity and reduced consumption volumes can contribute to extending the ranges.

Description

Energy productivity of the sector 'trade, commerce and services' describes how much Gross Value Added (GVA) is produced in this sector per GJ of energy.

Unit

€ GVA per GJ

Data

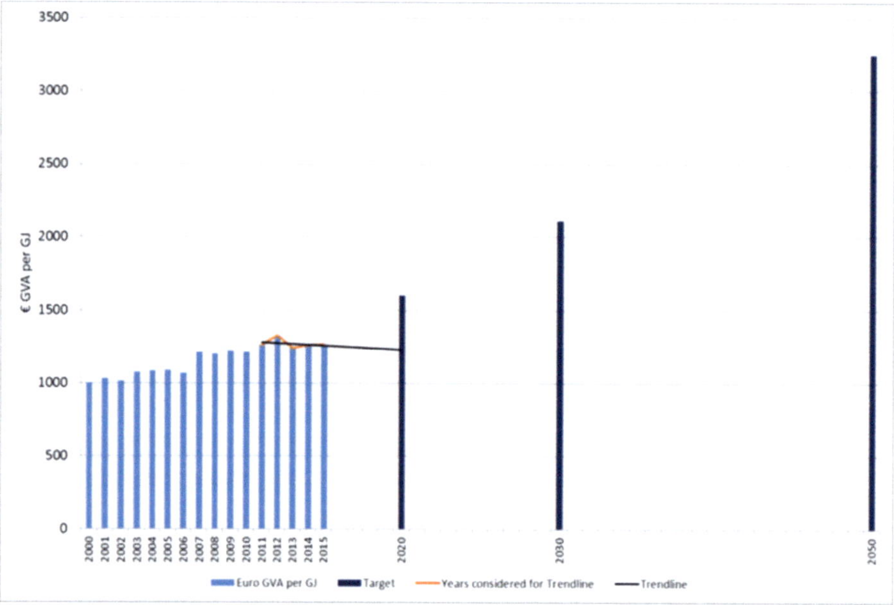

Figure 1: Final energy productivity of the sector 'trade, commerce and services' (temperature- and stock-adjusted) (own diagram based on [1])

Targets

2020: 1,602 €/GJ
2030: 2,111 €/GJ
2050: 3,251 €/GJ

Assessment

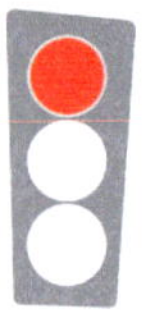

Value for the reference year (2010): 1,213 € GVA per GJ [1]

Latest available value (2015): 1,263 € GVA per GJ [1]

The trend calculated based on the previous 5 years (2011–2015) shows a decrease of the final energy productivity of 3 % by 2020 compared to the reference value (the average value over the years 2011 to 2015). The target value for 2020 requires an increase of 26 % compared to the reference value. Because the trend goes in a wrong direction not reaching the target of 2020, a red traffic light is assigned.

Comments Data	The data take into account final energy consumption and real GVA and are temperature- and stock-adjusted. Data are also available for primary energy consumption and thus for primary energy productivity.
Comments Targets	Objective of the Federal Government for the period until 2050 is an improvement of the overall energy productivity of 2.1 %/a. Starting in 2011, an increase in productivity of 2.1 %/a would lead to an increase by the factor 1.23 by 2020, by the factor 1.52 by 2030, and by the factor 2.3 by 2050 (each based on the value of 2010). Even though, according to [2], the increase in (primary) energy productivity of 1.65 %/a in the long-term average (1990–2010) was higher than the average growth rate of the GDP of 1.35 %/a, this minor difference is not sufficient for substantial reductions in consumption. Therefore [2] assumes an increase of the average growth rate of (primary) energy productivity to an average of 2.8 %/a between 2011 and 2030 and to an average of 2.2 %/a between 2031 and 2050. Starting in 2011, an increase in productivity of 2.8 %/a for the period 2011 to 2030 and of 2.2 % for the period 2031 to 2050 would lead to an increase by the factor 1.32 by 2020, by the factor 1.74 by 2030, and by the factor 2.68 by 2050 (each based on the value of 2010). These values, which are more ambitious than those of the Federal Government, which apply for the whole national economy, are also used for the target values of the sector 'Trade, commerce and services'. This also guarantees the consistency with the target values of other indicators, which are also taken from [2]. However, it must be noted that the growth rates given in [2] refer to primary energy productivity while the indicator refers to final energy productivity. Therefore the chosen target values might be more ambitious, since, for example, an increase of the share of solar energy in total energy production with otherwise identical boundary conditions leads to an increase in primary energy productivity, but not to an increase in final energy productivity. This results in the following target values (initial value 2010: 1,213 € per GJ) • 2020: 1,602 €/GJ (increase by the factor 1.32 compared to 2010) • 2030: 2,111 €/GJ (increase by the factor 1.74 compared to 2010) • 2050: 3,251 €/GJ (increase by the factor 2.68 compared to 2010)
Literature	[1] Bundesministerium für Wirtschaft und Energie: Fünfter Monitoring-Bericht zur Energiewende. Die Energie der Zukunft. Datenübersicht zum Fünften Monitoringbericht. November 2016. https://www.bmwi.de/Redaktion/DE/Artikel/Energie/monitoring-prozess.html, access 31 May 2017 [2] DLR – Deutsches Zentrum für Luft- und Raumfahrt; IfNE – Ingenieurbüro für neue Energien; IWES – Fraunhofer Institut für Windenergie und Energiesystemtechnik: Langfristszenarien und Strategien für den Ausbau der erneuerbaren Energien in Deutschland bei Berücksichtigung der Entwicklung in Europa und global. Schlussbericht BMU, FKZ 03MAP146, March 2012 http://www.dlr.de/dlr/Portaldata/1/Resources/bilder/portal/portal_2012_1/leitstudie2011_bf.pdf, access 29 June 2016

21. Energy-related Greenhouse Gas Emissions

<table>
<tr><td>Justification referring to Sustainability Concept</td><td>This indicator addresses the sustainability rule 'Sustainable use of the environment as a sink for waste and emissions': Based on the precautionary principle, the release of substances from the energy system should not exceed the carrying capacities of environmental media and ecosystems. Risks to the most sensitive parts of the overall system must be minimized in order to protect human health and to preserve vital natural regulating systems.</td></tr>
<tr><td>Description</td><td>Emissions of greenhouse gases (GHG) originating from the energy sector. They include the gases CO_2, CH_4 and N_2O in CO_2 equivalents (there are no other GHG emissions, such as HFC etc., resulting from the energy sector).</td></tr>
<tr><td>Unit</td><td>Million t of CO_2 equivalents</td></tr>
<tr><td>Data</td><td>

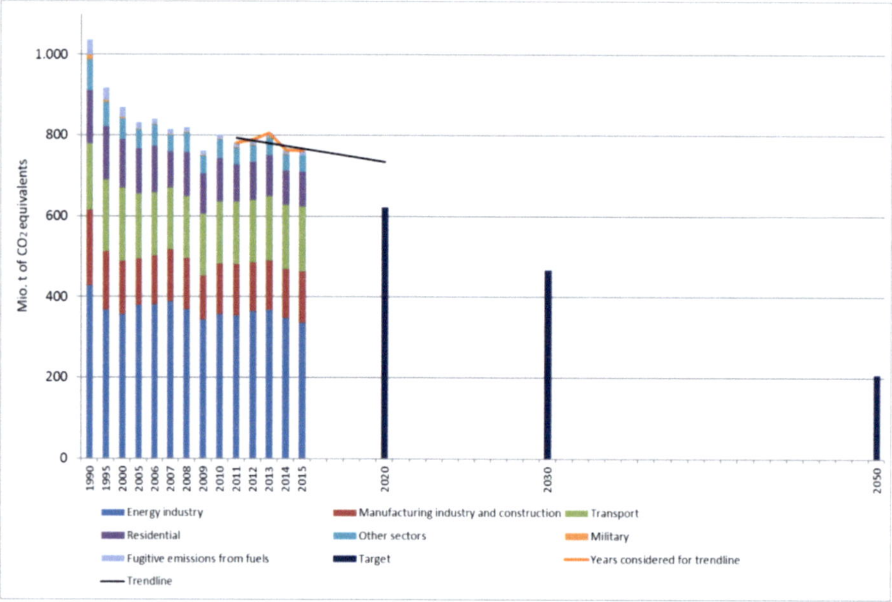

Figure 1: Greenhouse gas emissions resulting from the energy sector (own diagram based on [1])</td></tr>
<tr><td>Targets</td><td>2020: 622 million t of CO_2 equivalents
2030: 467 million t of CO_2 equivalents
2050: 207 million t of CO_2 equivalents</td></tr>
<tr><td>Assessment</td><td>

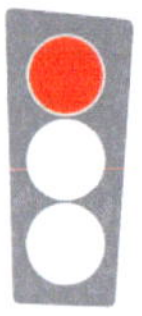

Value for the reference year (1990): 1,037 million t of CO_2 equivalents [1]
Latest available value (2015): 761 million t of CO_2 equivalents [1]
The trend calculated based on the previous 5 years (2011–2015) shows a decrease of the energy-related GHG emissions of 5.8 % until 2020 compared to the referenca value (the average value of the previous 5 years). The target value for 2020 requires a decrease of 20 % compared to the reference value. This results in a deviation of 71 % and a red traffic light, meaning that there is a high propability that the target of 2020 will be missed.</td></tr>
</table>

Comments Data	Only direct emissions are taken into consideration, because these data are published regularly. In addition, it has to be mentioned that GHG emissions due to imported products (emissions that occur in other countries – grey energy) are not considered. Therefore, a shift of production activities to other countries would reduce GHG emissions in Germany. Energy-related GHG emissions have a share of about 80 % for all the years considered. Energy-related GHG emissions per capita resulting from the energy sector amount to approx. 11 t for the year 1995 and to approx. 10 t per year for 2005 to 2015.
Comments Targets	According to the Paris Agreement net greenhouse gas emissions have to be cut back to zero between 2045 and 2060. The aim of the Federal Government is to reduce the GHG emissions by at least 40 % until 2020, 55 % until 2030 and 80 to 95 % until 2050 compared to 1990 levels (1,037 million t of CO_2 equivalents) [2]. This will result in the following target values: • 2020: 622 million t of CO_2 equivalents (reduction of 40 %) • 2030: 467 million t of CO_2 equivalents (reduction of 55 %) • 2050: 207 million t of CO_2 equivalents (reduction of 80 %)
Literature	[1] UBA – Umweltbundesamt: Nationale Trendtabellen für die deutsche Berichterstattung atmosphärischer Emissionen 1990–2015. https://www.umweltbundesamt.de/themen/klima-energie/treibhausgas-emissionen, access 14 February 2017 [2] UBA – Umweltbundesamt: Klimapolitische Ziele der Bundesregierung. Stand 03/2016. https://www.umweltbundesamt.de/sites/default/files/medien/384/bilder/dateien/4_tab_ziele-bundesreg_2016-10-07.pdf, access 09 Nov 2016

22. Energy-related Emissions of Acid-forming Gases

Justification referring to Sustainability Concept	This indicator addresses the sustainability rule *'Sustainable use of the environment as a sink for waste and emissions'*: Based on the precautionary principle, the release of substances from the energy system should not exceed the carrying capacities of environmental media and ecosystems. Risks to the most sensitive parts of the overall system must be minimized to protect human health and preserve vital natural regulating systems.
Description	Emissions of acid-forming air pollutants resulting from the energy sector include the gases SO_2, NO_x and NH_3 in SO_2 equivalents.
Unit	Million t of SO_2 equivalents
Data	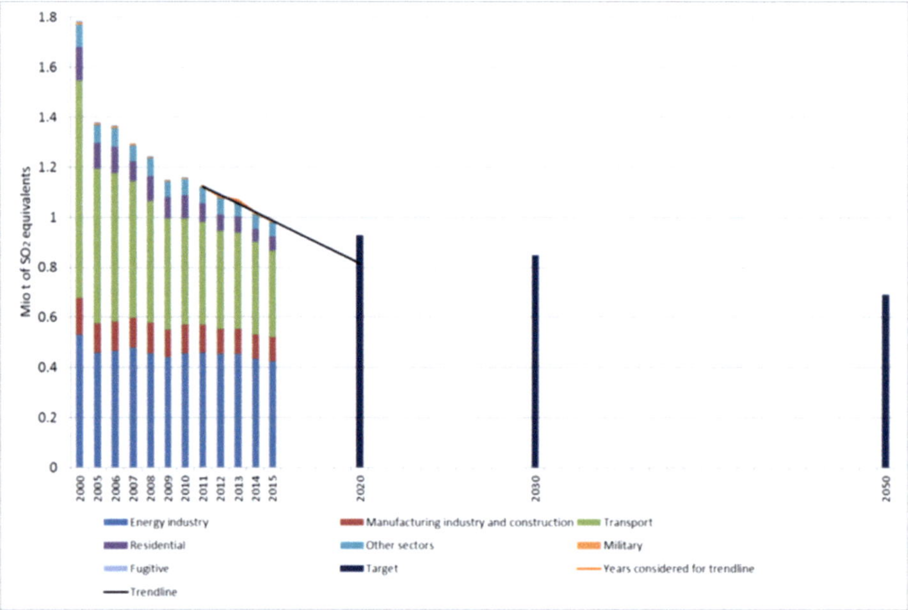 Figure 1: Energy-related emissions of acid-forming gases (own diagram based on [1]) The energy sector was responsible for about 29 % of all acid-forming emissions in 1995 and about 54 % in 2000. In the following years, the amount decreased to 43 % in 2015.
Targets	2020: 0.93 million t of SO_2 equivalents 2030: 0.85 million t of SO_2 equivalents 2050: 0.69 million t of SO_2 equivalents
Assessment	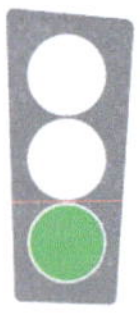 Value for the reference year (2005): 1.38 million t of SO_2 equivalents [1] Latest available value (2015): 0.98 million t of SO_2 equivalents [1] The target is a reduction of 17 % by 2020 compared to the reference value (the average value over the previous 5 years 2011 to 2015) and a reduction of 32.7 % compared to the reference year 2005. The trend calculated based on the previous 5 years shows a decrease of the emissions of acid-forming gases of 22.6 % until 2020 and a decrease of 41 % compared to 2005. So far, this indicator is on track to reach the target what results in a green traffic light.

Comments Data	Emissions of acid-forming gases from the energy sector do not only result from the energy industry itself, but also from energy use in the manufacturing sector, the transport sector, and the military sector, from other combustion sites as well as from energy-related fugitive emissions. Only direct emissions are taken into consideration. Values are given in SO_2 equivalents (with weighting factors of 0.7 for NO2, 1.88 for NH_3, and 1 for SO_2 [2]). Data are taken from [1].
Comments Targets	Directive 2001/81/EC of 23 October 2001 (NEC Directive) sets the national ceilings for emissions of the air pollutants sulphur dioxide (SO_2), nitrogen oxides (NO_x), ammonia (NH_3) and non-methane volatile organic compounds (NMVOC), which must not be exceeded after 2010. After this date, every Member State has to provide a national program to reduce these pollutant emissions. It has to prove the compliance with these emission ceilings by providing measures for emission control. Both the European Commission and the public have to be informed of the program. For Germany, this results in the following ceilings for acid-forming emissions in 2010 (total emissions): • SO_2: 0.52 million t • NO_x: 1.05 million t • NH_3: 0.55 million t The emissions for 2010 were: • SO_2: 0.41 million t • NO_x: 1.33 million t • NH_3: 0.68 million t In 2015, the emissions amounted to: • SO_2: 0.35 million t • NO_x: 1.86 million t • NH_3: 0.76 million t According to these values, emission ceilings for NO_x and NH_3 were exceeded in 2010 and 2013 [3]. In May 2012, the Parties to the Protocol agreed on an amendment of the Gothenburg Protocol. It establishes the emission reduction commitment (in %) for 2020 and the following years for the above-mentioned pollutants. The following reduction goals are set for Germany (reduction in % compared to the reference year 2005): SO_2: 21 %, NO_x: 39 %, NH_3: 5 % [4]. If these reduction goals are proportionally applied to energy-related emissions and the acidification potential is accordingly calculated (weighted averaging of SO_2, NO_x, und NH_3), resulting in a target value of 0.931 million t of SO_2 equivalents, which would require a reduction of energy-related emissions of 6 % compared to the values given for 2015 and 32.6 % to those for 2005. For 2050 a reduction of 50 % compared to the values given for 2005 is assumed. The value for 2030 is interpolated accordingly. This results in the following target values: • 2020: 0.93 million t (reduction of 32.7 % compared to 2005) • 2030: 0.85 million t (reduction of 38.5 % compared to 2005) • 2050: 0.69 million t (reduction of 50 % compared to 2005)

Literature	[1] UBA – Umweltbundesamt: Emissionen von Luftschadstoffen – Emissionsentwicklung 1990–2015 für klassische Luftschadstoffe. https://www.umweltbundesamt.de/themen/luft/emissionen-von-luftschadstoffen, access 27 April 2017 [2] IFEU – Institut für Energie - und Umweltforschung Heidelberg GmbH: Nachhaltiger Biogasausbau. Materialband E. Ökobilanzen. Im Rahmen des BMU-Forschungsvorhabens 'Optimierungen für einen nachhaltigen Ausbau der Biogaserzeugung und -nutzung in Deutschland', FKZ: 0327544. https://www.ifeu.de/landwirtschaft/pdf/BMU-Biogasprojekt%202008-Materialband%20E.pdf, access 23 Nov 2016 [3] UBA –Umweltbundesamt: Regelungen und Strategien. https://www.umweltbundesamt.de/themen/luft/regelungen-strategien, access 13 May 2015 [4] UBA – Umweltbundesamt: Strategien zur Emissionsminderung von Luftschadstoffen. https://www.umweltbundesamt.de/daten/luftbelastung/massnahmen-zur-emissionsminderung-von, access 13 May 2015

23. Energy-related Hazardous Solid Waste

Justification referring to Sustainability Concept	This indicator addresses the sustainability rule *'Sustainable use of the environment as a sink for waste and emissions'*: Based on the precautionary principle, the release of substances from the energy system should not exceed the carrying capacities of environmental media and ecosystems. Risks to the most sensitive parts of the overall system must be minimized in order to protect human health and to preserve vital natural regulating systems.
Description	Data on hazardous waste by different types of waste (waste classification) and different sectors (NACE classification) are given in the data bases of Eurostat [1] for the years 2004, 2006, 2008, 2010, 2012 and 2014. For the data presented here, the NACE category 'electricity, gas, steam, and air conditioning supply' has been evaluated. The waste categories shown in Table 1 were taken from the official statistics [1].
Unit	t
Data	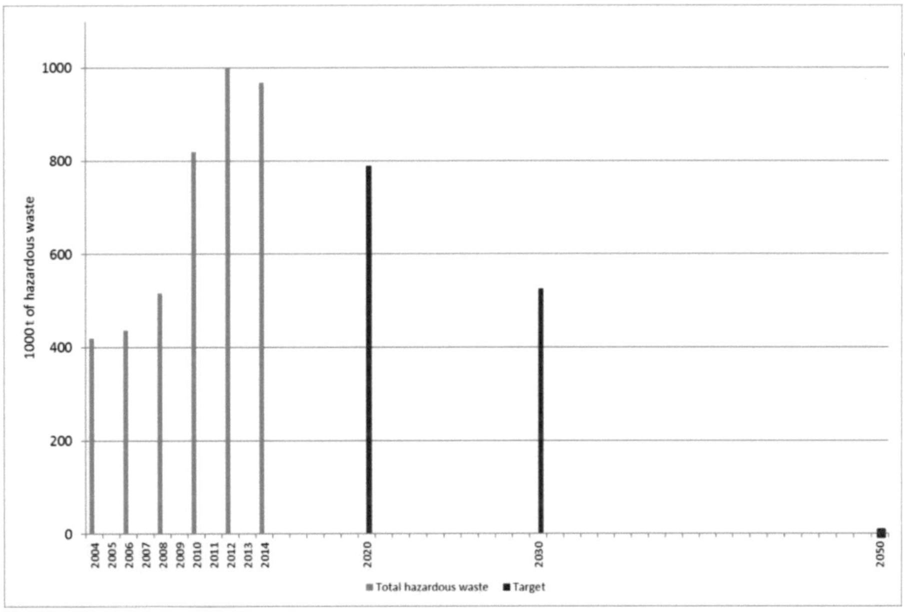 Figure 1: Amount of hazardous waste resulting from the sector 'electricity, gas, steam and air conditioning supply' (own diagram based on [1])
Targets	2020: 789 t 2030: 526 t 2050: 0 t
Assessment	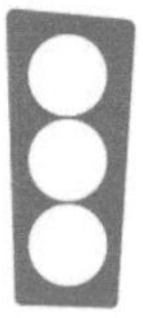 As can be seen in Figure 1, the amount of hazardous waste is increasing although it should decrease to result in no hazardous waste at all in 2050. Because data are only available for every second year, no trend line has been calculated.

Comments Data

The highest contribution to total hazardous waste results from mineral and solidified waste (774,155 t in 2014) followed by chemical and medical waste (135,509 t in 2014).

Detailed information about the composition of waste is given in Table 1.

Table 1: Amount of hazardous waste resulting from the sector 'electricity, gas, steam and air conditioning supply' [1]

Hazardous waste in tonnes per year						
	2004	2006	2008	2010	2012	2014
Chemical and medical	218,476	134,766	114,333	120,529	155,201	135,509
Recyclable	2,338	4,014	3,805	6,490	16,718	4,747
Equipment	5,259	6,934	10,862	21,803	33,092	27,838
Animal and vegetal						
Mixed ordinary	1,326	64	629	6,311	23,031	25,521
Common sludges						
Mineral and solidified	190,720	288,861	386,510	663,869	771,640	774,155
Total	418,681	435,836	516,245	819,399	999,682	967,770
There of combustion waste	186,565	284,952	377,373	32,741	20,641	41,685

Comments Targets

No waste ('0' in 2050) because all substances should be reused. The values for 2020 and 2030 were interpolated based on the values for 2014.

Literature

[1] Eurostat: Abfallaufkommen [env_wasgen]. http://ec.europa.eu/eurostat/statistics-explained/index.php/Waste_statistics/de, access 22 February 2017

24. Amount of High-level Radioactive Waste which has not been tranferred to a Safe Final Disposal Place

Justification referring to Sustainability Concept	This indicator addresses the sustainability rule *'Avoidance of technical risks with potentially catastrophic impacts'*: This includes risks due to, e.g., resource extraction, transport, energy generation and transmission, disposal, etc., associated with strong or potentially catastrophic impacts on humans or nature. Beyond the 'objective' measuring of such a risk, this also requires information about the public perception of existing or potential risks in order to identify the most relevant ones that need to be addressed most urgently.
Description	Radioactive waste, especially spent fuel elements and waste from reprocessing, involves risks and hazards for humans and the environment and must therefore be treated and disposed of adequately.
Unit	t HM (Heavy Metal)
Data	Latest available value (2013): 8,225 t HM for 2013 [1]
Targets	2050: 0 t HM
Assessment	 Since no time series are available for the assessment of this indicator it would have been methodologically correct to evaluate the indicator with a white traffic light. However, in this particular case we have made an exception. This is because of the total lack of high-level radioactive waste that is already in a safe final disposal place. It is expected that this situation will continue for a long time since the political process to identify suitable location has just been finished.
Comments Data	Only the amounts of spent fuel elements and waste from reprocessing have been taken into consideration. In the Federal Republic of Germany, radioactive waste originates from • the operation of nuclear power plants and research reactors, • the decommissioning of nuclear power plants, of experimental and demonstration reactors, research, as well as from research and training reactors for educational purposes, and other nuclear facilities, • uranium enrichment and fuel fabrication (nuclear industry), • basic and applied research, • the use of radioisotopes in other research institutions, universities, trade and industry companies, hospitals and medical practices, • other waste producers, such as the military sector, • future conditioning of spent fuel intended for direct disposal The 'Verzeichnis radioaktiver Abfälle' (list of radioactive waste) [1] provides an overview of the amount of radioactive waste and spent fuel in Germany (as of 31 December 2013) that has to be stored in an interim storage and finally moved to an adequate repository. The inventory of radioactive waste is subject to continuous change and it was said it will be updated every three years. However, an update for 2016 is not yet published.

	Several critics mention, that [1] exhibits gaps in counting radioactive waste in Germany. For the indicator, the amount of waste with the highest level of radioactivity is taken into consideration: spent fuel elements and waste from reprocessing.
Comments Targets	To reduce the risk of conatamination it is intended to transfer all high-level radioactive waste to a final storage place. Therefore, the target value should be '0'. However, in the following years the amount of radioactive waste from spent fuel and from reprocessing will rise due to an increase of spent fuels and due to the fact that no decision for a final storage place has been taken so far. Based on the fact that final storage will take an extremely long time, the share of total radioactive waste stored in a 'secure' interim storage would be a suitable indicator. In this case, a definition of 'secure interim storage' has to be provided.
Literature	[1] Bundesministerium für Umwelt, Naturschutz, Bau und Reaktorsicherheit, Kommission Lagerung hoch radioaktiver Abfallstoffe K-MAT13: Verzeichnis radioaktiver Abfälle. Bestand zum 31. Dezember 2013 und Prognose http://www.bundestag.de/blob/337852/7c57c8dc16bfc64f8ae86006964be6b2/kmat_13-data.pdf, access 9 April 2015 [2] Atommüllreport. http://www.atommuellreport.de/themen/atommuell/einzelansicht/radioaktive-abfaelle-mengenuebersicht.html, access 1 March 2017

25. Installed Capacity of Renewable Energy Power Plants

Justification referring to Sustainability Concept	This indicator addresses the sustainability rule *'Sustainable development of man-made, human, and knowledge capital'*: The man-made, human, and knowledge capital is to be developed in such a way that the economic capability, in terms of productive potential, can be maintained or improved. Aspects concerning the energy system include, e.g., maintenance or improvement of infrastructures related to man-made capital, education, training, and research related to human and knowledge capital.
Description	The precondition for producing more electricity from renewable resources is the installation of production capacities at renewable energy plants.
Unit	GW_p
Data	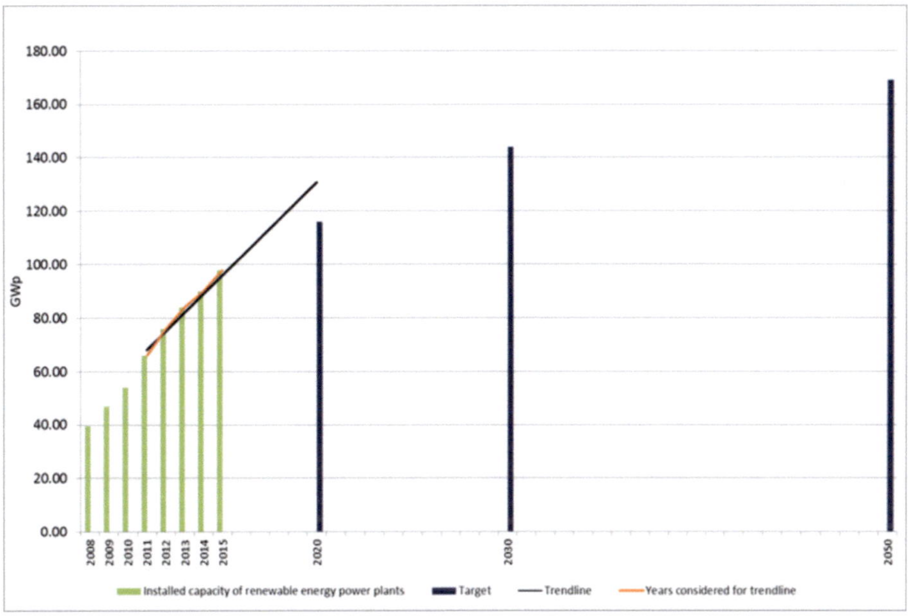 Figure 1: Installed capacity of renewable energy power plants in GW_p (own diagram based on [1, 2])
Targets	2020: 116 GW_p 2030: 144 GW_p 2050: 169 GW_p
Assessment	 Latest available value (2015): 98 GW_p [1] The trend calculated based on the previous 5 years (2011–2015) shows an increase of the installed capacity of 60 % until 2020 compared to the reference value (the average value over the years 2011 to 2015). The target value for 2020 requires an increase of only 42 % compared to the reference value. The expected overachievement of the target of 2020 results in a green traffic light.

Comments Data	The installed capacity includes PV power plants, wind power plants, the use of biomass for electricity production, geothermal energy and water power plants. Only plants situated in Germany are taken into consideration. The data do not give any information about the amount of electricity produced.
Comments Targets	Capacities to be installed in the future are geared to the future energy consumption, the targets of the energy transition (including the share of RE in future energy production), the mix of different types of plants and the security of the power supply (storage, reserve capacity). All these aspects are considered in [3]. Therefore, the respective target values for the years 2020, 2030 and 2050 were taken from this source. Capacities which are to be installed abroad in the relevant years for the power supply of Germany (import of power from RE) were not included.
Literature	[1] Bundesnetzagentur (BNetzA): Kraftwerksliste. Bundesnetzagentur. Stand November 2016. https://www.bundesnetzagentur.de/DE/Sachgebiete/Elektrizitaetund Gas/Unternehmen_Institutionen/Versorgungssicherheit/Erzeugungskapazitaeten/Kraftwerksliste/kraftwerksliste-node.html, access 09 February 2018 [2] Bundesministerium für Wirtschaft und Energie: Fünfter Monitoringbericht zur Energiewende. Die Energie der Zukunft. November 2016, p. 72 https://www.bmwi.de/Redaktion/DE/Publikationen/Energie/fuenfter-monitoring-bericht-energie-der-zukunft.pdf?__blob=publicationFile&v=23, access 22 February 2017 [3] DLR – Deutsches Zentrum für Luft- und Raumfahrt; IfNE – Ingenieurbüro für neue Energien; IWES – Fraunhofer Institut für Windenergie und Energiesystemtechnik: Langfristszenarien und Strategien für den Ausbau der erneuerbaren Energien in Deutschland bei Berücksichtigung der Entwicklung in Europa und global. Schlussbericht BMU, FKZ 03MAP146, March 2012, p. 311 http://www.dlr.de/dlr/Portaldata/1/Resources/bilder/portal/portal_2012_1/leitstudie2011_bf.pdf, access 29 June 2016

26. Number of University Graduates in the Field of Energy Sciences

Justification referring to Sustainability Concept	The indicator addresses the sustainability rule *'Sustainable development of man-made, human and knowledge capital'*: The Man-made, human and knowledge capital is to be developed in such a way that the economic capability, in terms of productive potential, can be maintained or improved. Aspects concerning the energy system include, e.g., maintenance or improvement of infrastructures in the context of man-made capital, education, training, and research in the context of human and knowledge capital.
Description	Total number of graduates per year and per subject in the courses 'Energy engineering (excluding electrical engineering since the focus here is not on energy production technology but on electrical appliances)', 'Electronic energy engineering', and 'Renewable energies'.
Unit	Number of university graduates per year
Data	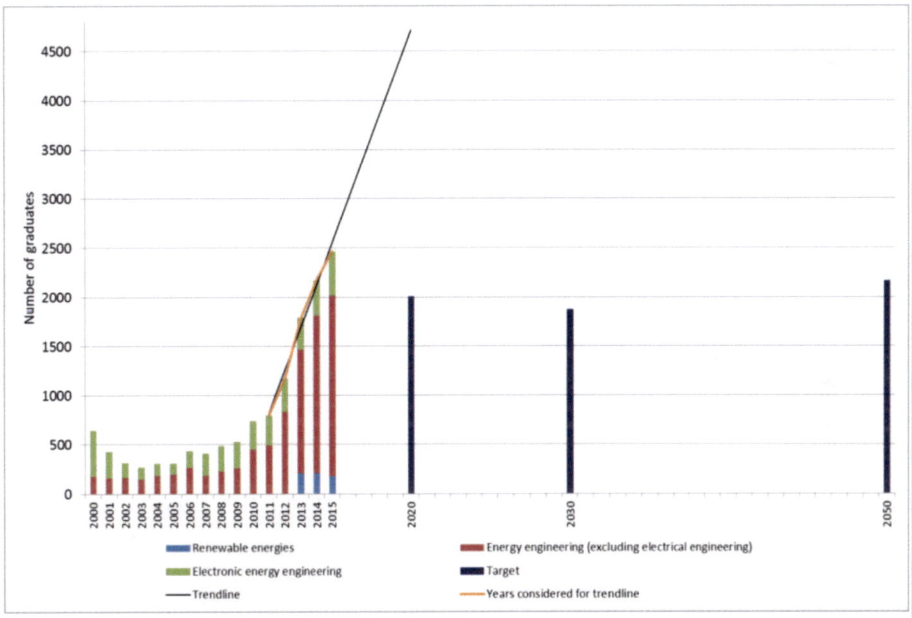 Figure 1: Number of university graduates in the field of energy sciences (own diagram based on [1])
Targets	2020: 2,702 graduates 2030: 2,516 graduates 2050: 2,919 graduates
Assessment	 Latest available value (2015): 2,464 university graduates [1] The trend calculated based on the previous 5 years (2011–2015) shows an increase of the graduates of about 163 % until 2020 compared to the reference value (the average value over the years 2011 to 2015). The minimum target value for 2020 requires just an increase of 61 % compared to the reference value. Thus, a green traffic light is given.

Comments Data	Data for the course 'Renewable energies' are only available since 2013. However, to monitor the graduates in energy and energy efficiency courses in general, a special recording structure is recommended. It should cover all graduates in energy-related subjects. In [2], about 300 degree courses with specialization in RE are listed. The corresponding numbers of graduates in these courses are not known and only few of them are included in the data of Figure 1.
Comments Targets	The following dependencies are assumed to derive the target values: • The number of jobs in the field of RE depends on the investments in this sector. • The demand for qualified employees in turn influences the choice of subjects and thus also the number of academic degrees. Therefore we can assume that the number of final degrees in the courses 'Energy engineering (excluding Electrical engineering)', 'Electronic energy engineering', and 'Renewable energies' will develop proportionally to the number of jobs in the field of RE (see also Indicator 7). The basic value for the calculation is the previous year with available data (2015) with 330,000 jobs (see Indicator 7) in the field of RE and 2,424 final degrees in the field of RE.
Literature	[1] DESTATIS – Statistisches Bundesamt: GENESIS-Online Datenbank. https://www-genesis.destatis.de/genesis/online/data;jsessionid=4B5CA2EE9B7FD908DA4199F1BADA9DF6.tomcat_GO_2_2?operation=abruftabelleAbrufen&selectionname=21321-0003&levelindex=1&levelid=1430750648551&index=3, access 14 February 2017 [2] Studium Erneuerbare Energien – Das Informationsportal zum Studium im Bereich erneuerbare Energien: Übersicht Studiengänge mit vollständiger Ausrichtung auf Erneuerbaren Energien. http://www.studium-erneuerbare-energien.de/ubersicht-studiengange-mit-vollstandiger-ausrichtung-auf-erneuerbaren-energien.php, access 7 May 2015

27. Federal Expenditures for Energy Research

Justification referring to Sustainability Concept	This indicator addresses the sustainability rule *'Sustainable development of man-made, human, and knowledge capital'*: The man-made, human, and knowledge capital is to be developed in such a way that the economic capability, in terms of productive potential, can be maintained or improved. Aspects concerning the energy system include, e.g., maintenance or improvement of infrastructures related to man-made capital, education, training, and research related to human and knowledge capital.
Description	Expenditures for energy research by Federal Government and Federal States
Unit	Million €

Data

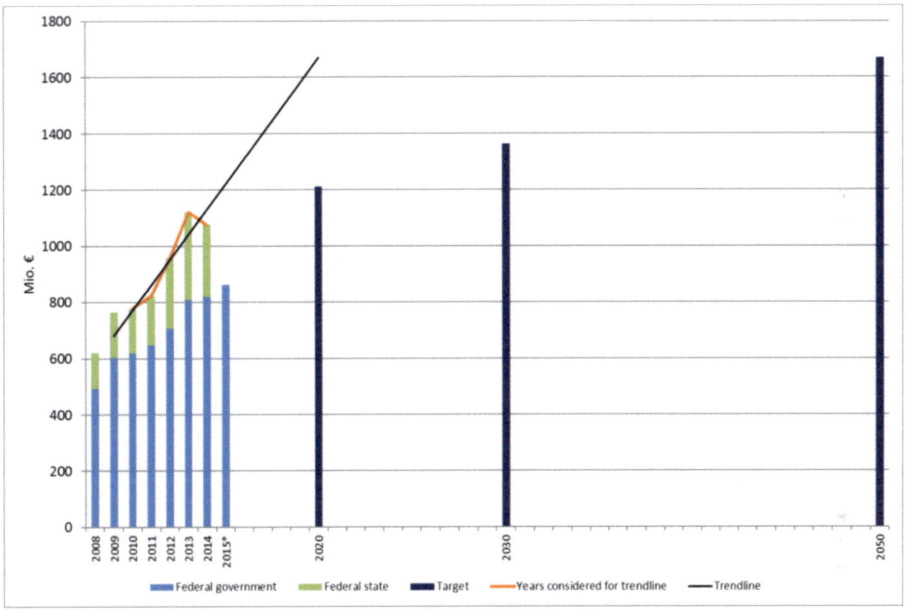

Figure 1: Federal expenditures for energy research (own diagram based on [1-2])
*No information available about the expenditures of the Federal States for 2015

Targets	2020: 1,212 million € 2030: 1,365 million € 2050: 1,670 million €
Assessment	Latest available value (2014): 1,076 million € [1] The years 2010 to 2014 were used to calculate the trend since there is no information available about the expenditures of the Federal States in 2015. The extrapolation of the trend until 2020 shows significantly higher values than the target value requires, which results in a green traffic light.

Comments Data	Up to now, there is no information available about the expenditures of the Federal States for 2015. The indicator presented here provides no data on private and university research or the research of the Helmholtz Association (HGF). Further detailed information and data are available on, e.g., energy conversion, energy distribution and energy consumption. These data show, among other things, a strong research on renewable energy at the level of the Federal States (see [2]). Additional research funding in the field of nuclear energy research in Germany is based on the EURATOM treaty. We propose to monitor the research expenditures in the field of energy for all subcategories for each Federal State (spatially differentiated) on an annual basis (in line with the Federal Government's expenditures).
Comments Targets	The fact that a number of sustainability goals for the energy transition are not met shows that there are some issues in need of further action. However, these actions are not clearly defined. So it can be derived that further research in this field is necessary and that its scope has to be increased. The research spending in Germany in relation to the GDP and the research spending of the country with the highest value in this category are used as reference point for future expenditures. As a country that is poor in natural resources, Germany should aim for the best with its investments in research. This is South Korea with 4.29 % in 2014 [3]. In our opinion, the goal for Germany should be to increase its research spending from 2.88 % to date [3] to the mentioned 4.29 % of the GDP in 2050. This corresponds to an increase by the factor 1.49. The research spending for the energy sector should increase by the same factor to make sure that the share of energy research in total research spending remains the same. Here it has to be mentioned that these data are real values. Therefore, the values have to be adjusted according to the inflation rate. The values for 2020 and 2030 were interpolated from the value for 2014.
Literature	[1] BMWi – Bundesministerium für Wirtschaft und Energie: Bundesbericht Energieforschung 2016. Forschungsförderung für die Energiewende. 2017, p 42, https://www.ptj.de/lw_resource/datapool/_items/item_7306/bundesbericht-energieforschung-2016.pdf, access: 7 February 2017 [2] BMWi – Bundesministerium für Bildung und Forschung: Ausgaben für Energieforschung in Deutschland durch den Bund in den Jahren 1991 bis 2015. https://de.statista.com/statistik/daten/studie/152800/umfrage/bund---ausgaben-fuer-energieforschung-seit-1991/, access 7 February 2017 [3] Statista. Ausgaben für Forschung und Entwicklung in Prozent des BIP in ausgewählten Ländern im Jahr 2014. https://de.statista.com/statistik/daten/studie/158150/umfrage/ausgaben-fuer-forschung-und-entwicklung-2008/, access 7 February 2017

28. Number of German Patents in the Field of Renewable Energy and Energy Efficiency

Justification referring to Sustainability Concept	This indicator addresses the sustainability rule *'Sustainable development of man-made, human and knowledge capital'*: The man-made, human and knowledge capital is to be developed in such a way that the economic capability, in terms of productive potential, can be maintained or improved. Aspects concerning the energy system include, e.g., maintenance or improvement of infrastructures in the context of man-made capital, education, training, and research in the context of human and knowledge capital.
Description	Number of patents with German participation based on OECD.StatExtracts [1].
Unit	Patents per year
Data	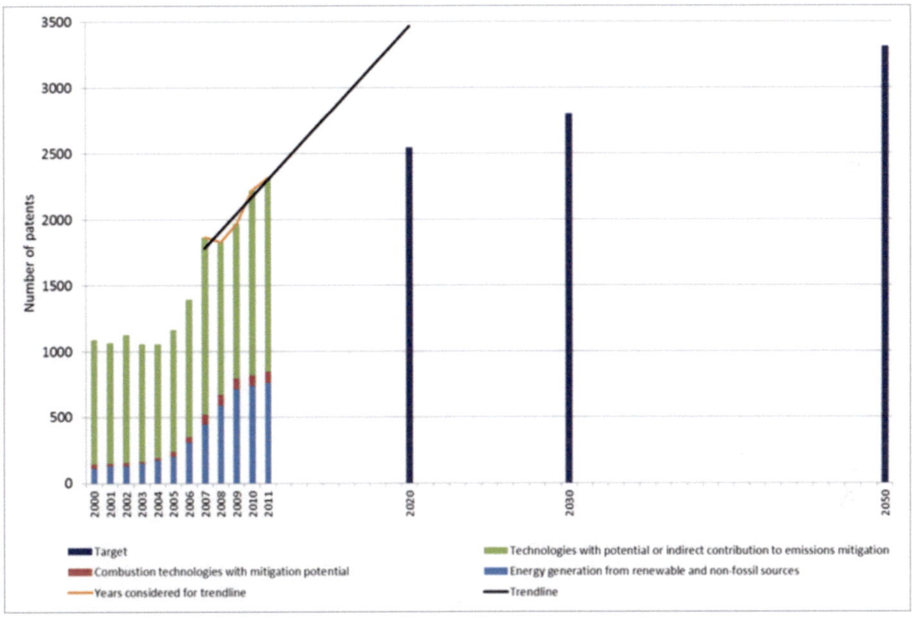 Figure 1: Evolution of the number of patents in Germany in the field of renewable energies and energy efficiency (own diagram based on [1])
Targets	2020: 2,580 patents 2030: 2,874 patents 2050: 3,459 patents
Assessment	 Latest available value (2011): 2,317 patents [1] The trend calculated based on the previous 5 years (2007–2011) shows an increase of the patents of 70 % until 2020 compared to the reference year (the average value over the years 2007 to 2011). The target value for 2020 requires an increase of only 26 % compared to the average measured values for the years 2007 to 2011. A green traffic light is given since it is likely that the target of 2020 will met.

Comments Data

From the OECD patent data base, the following items have been used:

- Patents application to the EPO (European Patent Office)
- Inventor's/inventors' country/countries of reference
- Technology domain: environment-related technologies with the subgroups shown in Table 1.

Table 1: Subgroups taken into consideration according to OECD patent statistics [1]

Energy generation from renewable and non-fossil sources		
	Renewable energy generation	
		Wind energy
		Solar thermal energy
		Solar photovoltaic (PV) energy
		Solar thermal-PV hybrids
		Geothermal energy
		Marine energy (excluding tidal)
		Hydro energy – tidal, stream or damless
		Hydro energy – conventional
	Energy generation from fuels of non-fossil origin	
		Biofuels
		Fuel from waste (e.g., methane)
Combustion technologies with mitigation potential (e.g., using fossil fuels, biomass, waste, etc.)		
	Technologies for improved output efficiency (combined combustion)	
		Heat utilization in combustion or incineration of waste
		Combined heat and power (CHP)
		Combined cycles (incl. CCPP, CCGT, IGCC, IGCC+CCS)
	Technologies for improved input efficiency (efficient combustion or heat usage)	
Technologies specific to climate change mitigation		
	Capture, storage, sequestration or disposal of greenhouse gases	
		CO_2 capture and storage (CCS)
		Capture and disposal of greenhouse gases other than carbon dioxide (incl. N_2O, CH_4, PFC, HFC, SF_6)
Technologies with potential or indirect contribution to emissions mitigation		
	Energy storage	
	Hydrogen production (from non-carbon sources), distribution and storage	
	Fuel cells	
Emissions abatement and fuel efficiency in transportation		
	Technologies specific to propulsion using internal combustion engine (ICE) (e.g., conventional petrol/diesel vehicle, hybrid vehicle with ICE)	
	Technologies specific to propulsion using internal combustion engine (ICE) (e.g., conventional petrol/diesel vehicle, hybrid vehicle with ICE)	Integrated emissions control (NOX, CO, HC, PM)
		Post-combustion emissions control (NOX, CO, HC, PM)
	Technologies specific to propulsion using electric motor (e.g., electric vehicle, hybrid vehicle)	

<table>
<tr><td></td><td>
<table>
<tr><td colspan="2">Energy efficiency in buildings and lighting</td></tr>
<tr><td></td><td>Insulation (incl. thermal insulation, double-glazing)</td></tr>
<tr><td></td><td>Heating (incl. water and space heating, air-conditioning)</td></tr>
<tr><td></td><td>Lighting (incl. CFL, LED)</td></tr>
</table>
When a patent was invented by several inventors from different countries, the respective contribution of each country is taken into account. This is done in order to avoid multiple counting of such patents.
Data for the years 2012 and 2013 are currently not available. Additionally, the analyses of patents are associated with certain disadvantages (e.g., it remains unclear whether the patented inventions are brought to market).
</td></tr>
<tr><td>Comments Targets</td><td>
A linear relation between research spending in the energy sector (see Indicator 27) and registered patents in the fields of RE and energy efficiency is presumed. It is also assumed that research spending will rise until 2050 by the factor 1.49 compared to 2014.
This value is also taken as basis for the patents, whereas the initial value is the number of patent applications for 2011 with 2,317 patents (the most recent year with available data). The values for 2020 and 2030 were interpolated accordingly.
</td></tr>
<tr><td>Literature</td><td>[1] OECD.StatExtracts: Environment-related patents by technology. http://stats.oecd.org/, access 7 May 2015</td></tr>
</table>

29. Number of Start-ups in the Renewable Energy and Energy Efficiency Sector

Justification referring to Sustainability Concept	This indicator addresses the sustainability rule *'Sustainable development of man-made, human and knowledge capital'*: Man-made, human and knowledge capital has to be developed in such a way that society's productive potential, particularly its economic capability, can be maintained or improved. Aspects concerning the energy system include, e.g., maintenance or improvement of infrastructures in the context of man-made capital, education, training, and research related to human and knowledge capital.
Description	Start-ups can be considered as drivers of innovation. Looking at the German energy transition, start-ups in the renewable energy and energy efficiency sector play an important role. Here, innovative products and processes are explored and produced or existing technologies are advanced [1, p. 83].
Unit	Number of start-ups
Data	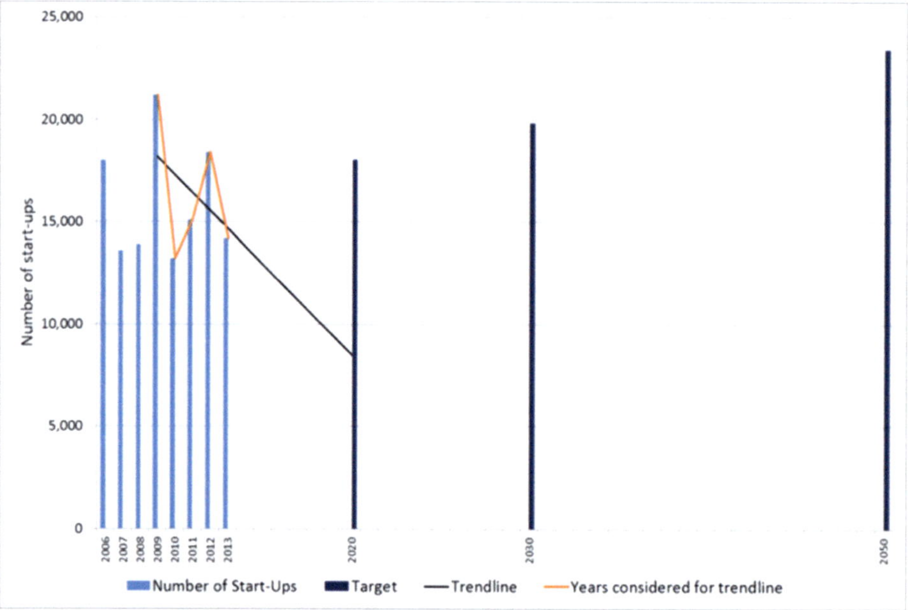 Figure 1: Number of start-ups in the renewable energy and energy efficiency sector (own diagram based on [2-3])
Targets	2020: 18,288 start-ups 2030: 20,363 start-ups 2050: 24,515 start-ups
Assessment	 Latest available value (2013): 14,200 start-ups [3] The trend calculated based on the previous 5 years (2009–2013) shows a decrease in the number of start-ups of approx. 48 % by 2020 compared to the reference value (the average value over the years 2009 to 2013). However, the target value for 2020 is about 11 % higher than the reference value. A red traffic light is assigned since it is not likely that the target of 2020 will be met.

Comments Data	The data is derived from a database containing information on 1.3 million active businesses in Germany [2]. The sample size was 5,000. The data was classified according to the 'Environmental Goods and Services Sector' method. Thus, the start-ups could be assigned to eight targeted areas of the green economy. These areas include: climate protection, renewable energies, energy efficiency, emission prevention, recycling economy, resource efficiency, renewable resources, and biodiversity. Only the businesses in the renewable energy and energy efficiency sector were considered in the evaluation (adjusted for duplicates/businesses active in both sectors). The numbers for start-ups presented above differ significantly from those presented in [1] (which are based on [4]). One reason is that [4] uses a more conservative method to ascribe start-ups to the RE sector: it is based on a keyword search within the company name and description (inside the database). The Borderstep Institute uses individual internet-based research to classify the businesses within the sample. Both are based on the Creditreform database.
Comments Targets	It is assumed that the number of start-ups is interrelated to the number of registered patents in the renewable energy and energy efficiency sector (Indicator 28). The number of newly registered patents, in turn, is assumed to be dependent on the amount of spending on energy research (Indicator 27). Research expenditure is assumed to increase by a factor of 1.49 by the year 2050 compared to 2014. The same value is assumed for start-ups, with the initial value being the average of the number of start-ups over the previous 5 years for which data are available (16,420 start-ups). The values for the years 2020 and 2030 were interpolated accordingly.
Literature	[1] BMWi – Bundesministerium für Wirtschaft und Energie: Ein gutes Stück Arbeit – Die Energie der Zukunft. Erster Fortschrittsbericht zur Energiewende, Berlin, December 2014. http://www.bmwi.de/Redaktion/DE/Publikationen/Energie/fortschrittsbericht.pdf?__blob=publicationFile&v=11, access 30 April 2015 [2] Borderstep Insitute: Green Economy Gründungsmonitor 2014 – Grüne Wirtschaft als Gründungs- und Beschäftigungsmotor in Deutschland. 2015. http://www.borderstep.de/wp-content/uploads/2015/05/Green_Economy_Gruendungsmonitor_20141.pdf, access 27 May 2015 [3] Borderstep Institute (own request in May 2015) [4] ZEW – Zentrum für Europäische Wirtschaftsforschung: Potenziale und Hemmnisse von Unternehmensgründungen im Vollzug der Energiewende. Studie im Auftrag des Bundesministeriums für Wirtschaft und Energie, 2014. https://www.bmwi.de/Redaktion/DE/Publikationen/Studien/potenziale-und-hemmnisse-von-unternehmensgruendungen-im-vollzug-der-energiewende.html, access 12 May 2015

30. Added Value Creation from the Renewable Energy Sector

<table>
<tr><td>Justification referring to Sustainability Concept</td><td>This indicator addresses the sustainability rule 'Sustainable development of man-made, human and knowledge capital': Man-made, human and knowledge capital has to be developed in such a way that society's productive potential, particularly its economic capability, can be maintained or improved. Aspects concerning the energy system include, e.g., maintenance or improvement of infrastructures in the context of man-made capital, education, training, and research related to human and knowledge capital.</td></tr>
<tr><td>Description</td><td>The indicator summarizes the creation of gross value added (GVA) at the national level through renewable energies.</td></tr>
<tr><td>Unit</td><td>€</td></tr>
<tr><td>Data</td><td>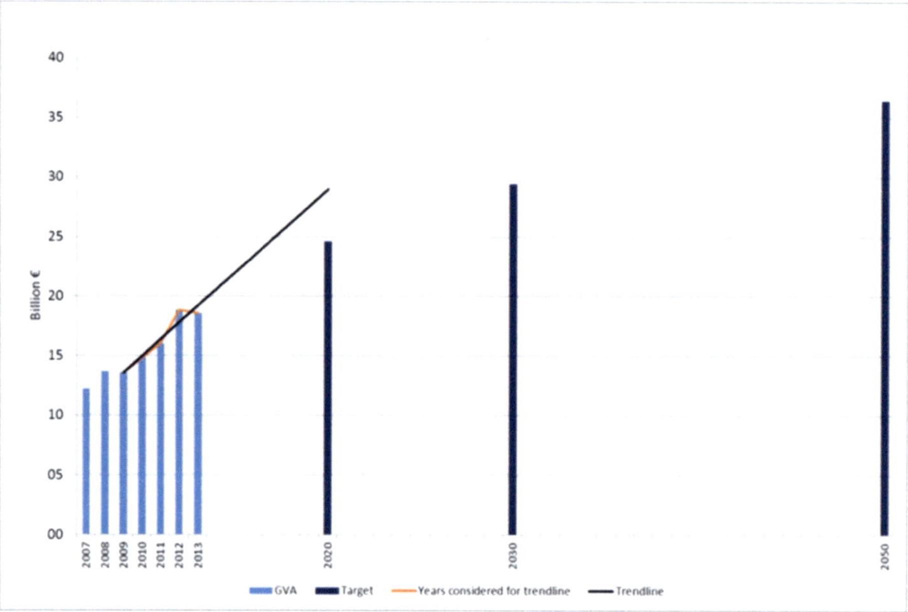

Figure 1: Evolution of the creation of gross value added (GVA) from the renewable energy sector (own diagram based on [1-3])</td></tr>
<tr><td>Targets</td><td>2020: 24.6 billion €
2030: 29.4 billion €
2050: 36.4 billion €</td></tr>
<tr><td>Assessment</td><td>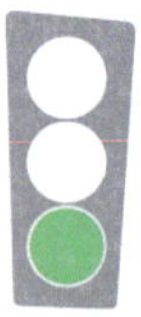
Latest available value (2013): 18.6 billion € GVA [1]
The trend calculated based on the previous 5 years (2009–2013) shows a strong increase in gross value added until 2020 compared to the reference value (the average value over the years 2009 to 2013). Given this current trend, the value for 2020 (29.0 billion €) would thus exceed the target value for 2020 (24.6 billion €) by more than 17 %. A green traffic light is given since the indicator is on track to meet the target of 2020.</td></tr>
</table>

Comments Data	'Based on the rules developed [by Prognos], gross value added and employment can now be disclosed via the newly delineated cross-sector energy industry. This is used to allocate all classic 2-digit industry sectors proportionately to the energy sector. The data basis consists of the latest available data from the German gross domestic product calculation (detailed annual results) in the national accounts of the Federal Statistical Office (posted: 26 May 2014)' (own translation) [1, p. 52]. The following steps were undertaken to calculate the data points: (1) For all German economic branches, the respective shares of gross value added attributable to energy were identified. (2) Then, from these shares the percentages clearly attributable to renewable energy and conventional energy were established, along with the remaining share of energies that are not clearly distinguishable ('mixed'). (3) Finally, from this percentage of 'mixed' energies – as an approximation [4, p. 55, footnote] – the share of renewables was set equal to the electricity mix during the respective year [2]. (4) Besides the value for the reference year 2011 published by Prognos, additional time series were then generated using the annual gross value added data from [3], assuming the shares used in steps 1 and 2 remain approximately the same as set out for the year 2011 by Prognos [1, p. 57]. The formula used for the time series: gross added value (GVA) from renewable energies (RE) in the year x = *Σ over all economic branches* { GVA of economic branch in year x * [(GVA share attributed to energy * share of RE in GVA) + (GVA share attributed to energy * share of 'mixed' energy in GVA * share of RE in electricity mix in year x)] } Alternatively to the data and method by Prognos selected above, the 'bottom-up' modeling of IÖW could be applied [4 - 6]: The value creation at the national level through renewable energies (RE) would then be derived from the company profits, taxes at municipal and national levels, and the income from wages, which are all generated through four value creation stages in several technology-dependent value creation chains (including systems engineering/production, planning and installation of RE, site operation and maintenance, system operation). Thus, according to IÖW, the modelled value added through RE in Germany in 2012 approximates to 29.8 billion € (direct 18.9 billion € + indirect 10.9 billion € value added) [4, p. 208]. For the year 2012, the value added of 18.9 billion € direct volume added derived from the IÖW approach corresponds very well with the value added of 18.8 billion € calculated from the Prognos approach. Nevertheless, a valid comparison cannot be made with data of only one year and given the very diverse nature of the two approaches. Since the approach by Prognos allows to establish time series and, therefore, to make an assessment – this one was selected here.

Comments Targets

However, the Prognos approach cannot be used to derive appropriate target values, since the shares provided for each economic branch cannot be extrapolated to the (far) future – mainly because of potential dynnamic shifts in the future development of the branches. Therefore, the IÖW approach, which is solidly grounded in numerous well-developed technology value chains, seems suitable for target derivation.

According to [6], the creation of value added for RE breaks down as follows:

- Production: 36 %; Planning & Installation: 13 % – together: 49 %
- Operation & Management: 24 %; System Operation: 27 % – together 51 %

These shares are applied to the value added of the reference year 2012 (18.8 billion €). That means for the future development that 49 % of value added creation varies with the change of investments (taken from [7]) and 51% of future added value creation varies with the change of electricity and heat produced (taken from [7, 8]). This results in the following target values:

	Investments [7] billion €	**Electricity and heat produced from RE [7, 8] PJ**	**Value added creation billion €**
2012	17.42	1,008	18.8
2020	18.43	1,557	24.6
2030	17.16	2,135	29.4
2050	19.91	2,716	36.4

Literature

[1] Prognos: Wertschöpfungs- und Beschäftigungseffekte der Energiewirtschaft. Schlussbericht. Studie im Auftrag des Bundesministeriums für Wirtschaft und Energie BMWi. March 2015

[2] AGEB – AG Energiebilanzen e.V.: Strommix. Stromerzeugung nach Energieträgern 1990–2015 (status: 28 January 2016) http://www.ag-energiebilanzen.de/index.php?article_id=29&fileName=20160128_brd_stromerzeugung1990-2015.pdf, access 9 June 2016

[3] Statistisches Bundesamt: Volkswirtschaftliche Gesamtrechnungen. Inlandsproduktsberechnung. Detaillierte Jahresergebnisse. Fachserie 18 Reihe 1.4, 2016

[4] IÖW – Institut für ökologische Wirtschaftsforschung: Wertschöpfung durch Erneuerbare Energien. Ermittlung der Effekte auf Länder- und Bundesebene. December 2015

[5] Renewable Energies Agency: Value Creation for Local Communities through Renewable Energies, Renews Special, 2010. https://www.germany.info/contentblob/3097466/Daten/1196468/RenewsSpecial_DD.pdf, access 31 May 2017

[6] Prahl, A. (Institut für ökologische Wirtschaftsforschung – IÖW): Renewable energies' impact on value added and employment in Germany – Model results for 2012, Conference presentation, 2014. https://www.ioew.de/uploads/tx_ukioewdb/Prahl_Andreas_Renewable_energies_and_value_added_Fukushima_Febr-2014.pdf, access 31 May 2017

	[7] DLR – Deutsches Zentrum für Luft- und Raumfahrt; IfNE – Ingenieurbüro für neue Energien; IWES – Fraunhofer Institut für Windenergie und Energiesystemtechnik: Langfristszenarien und Strategien für den Ausbau der erneuerbaren Energien in Deutschland bei Berücksichtigung der Entwicklung in Europa und global. Schlussbe-richt BMU – FKZ 03MAP146, March 2012 http://www.dlr.de/dlr/Portal-data/1/Resources/bilder/portal/portal_2012_1/leitstudie2011_bf.pdf, access 29 June 2016 [8] BDEW – Bundesverband der Energie- und Wasserwirtschaft e.V.: Erneuerbare Energien und das EEG: Zahlen, Fakten, Grafiken, 2013. https://www.bdew.de/inter-net.nsf/id/17DF3FA36BF264EBC1257B0A003EE8B8/$file/Foliensatz_Energie-Info-EE-und-das-EEG2013_31.01.2013.pdf, access 9 June 2016

31. Added Value Creation from Energy Efficiency Measures in Households

Justification referring to Sustainability Concept	The indicator addresses the sustainability rule *'Sustainable development of man-made, human and knowledge capital'*: Man-made, human and knowledge capital has to be developed in such a way that society's productive potential, particularly its economic capability, can be maintained or improved. Aspects concerning the energy system include, e.g., maintenance or improvement of infrastructures in the context of man-made capital, education, training and research related to human and knowledge capital.
Description	The indicator summarizes the value creation at the national level through energetic refurbishment measures in the form of company profits, taxes at municipal and national levels, and the income from wages, which are all generated through four value creation stages in several value creation chains (including building services, window replacement, insulation of basement ceiling, insulation of top floor ceiling, external wall insulation).
Unit	€
Data	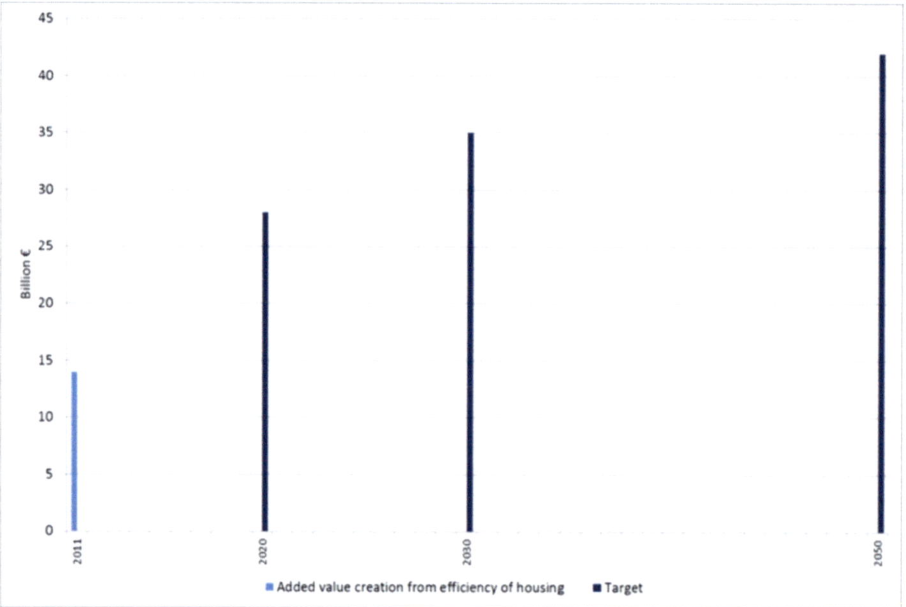 Figure 1: Total value added through the energetic refurbishment of houses (own diagram based on [1])
Targets	2020: 28 billion € 2030: 35 billion € 2050: 42 billion €
Assessment	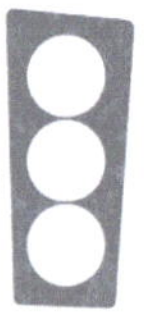 The total value added through the energetic refurbishment of houses in Germany in 2011 is approx. 14 billion € [1] There is no assessment possible because values are only available for 2011.

Comments Data	The total value from 2011 (approx. 14 billion €) corresponds to 0.5 % of GDP and 278,000 full-time jobs. Value creation through energy efficiency measures has not been reported in official studys yet (besides the IÖW modeling efforts). In the light of the high significance of energy efficiency measures in the German energy system, an assessment of value creation from energy efficiency is desirable in order to be able to quantify economic aspects related to those measures as well. Approach adopted (IÖW) [1]: The national bottom-up value creation is calculated as the sum of • net profits of enterprises involved • net incomes of employees involved • taxes and other charges paid to federal, state and local governments. The municipal share of the value creation is calculated for different value creation chains of energy efficiency measures. IÖW compiled its study using the defined value chains, cost structures, and reference processes, by determining the local added value and employment effects, and finally extrapolating to the German national level. The specific methodical steps were: (1) Selection of reference buildings (2) Detailed analysis of the current status (3) Selection of refurbishment measures (4) Energy requirements for the refurbishment; development of a package of measures (5) Energy demand calculations for the reference buildings (6) Development of a method for the creation of value-added effects (7) Determination of the cost structures of value chains for the individual measures and for the packages The following simplifications were made: For all buildings in the building stock, the specific costs of the refurbishment measures were adopted corresponding to those of the three reference buildings. Only refurbishment measures with the available cost structures were considered. For 2011, some assumptions had to be made about the refurbishment activities since no exact figures were available.

Comments Targets	The refurbishment rate is assumed to continuously increase from presently (2010) 1 % per year to 2 % per year by 2020 (see also [2]), in each case relating to the building stock to be refurbished in 2010. Furthermore, it is assumed that the rate continues to increase to 3 % per year by 2040 and this value remains constant until the year 2050. Under these assumptions, almost the entire building stock to be refurbished will be renovated by 2050. Furthermore, it is assumed that 'value added creation' increases in proportion to the refurbishment rate, i.e., a doubling of the refurbishment rate also leads to a doubling of 'value added creation' (deviating values may result depending on the depth of refurbishment). From these assumptions, the following target values have been derived:

	Refurbishment rate % per year	**Value added creation billion €**
2011	1	14
Since 2020	2	28
Since 2030	2.5	35
2050 (since 2040)	3	42

Literature

[1] IÖW – Institut für ökologische Wirtschaftsforschung; Ecofys: Kommunale Wertschöpfungseffekte durch energetische Gebäudesanierung (KoWeG). 2014 https://www.ioew.de/projekt/kommunale_wertschoepfungseffekte_durch_energetische_gebaeudesanierung/, access 01 June 2017

[2] DLR – Deutsches Zentrum für Luft- und Raumfahrt; IfNE – Ingenieurbüro für neue Energien; IWES – Fraunhofer Institut für Windenergie und Energiesystemtechnik: Langfristszenarien und Strategien für den Ausbau der erneuerbaren Energien in Deutschland bei Berücksichtigung der Entwicklung in Europa und global. Schlussbericht BMU – FKZ 03MAP146, March 2012 http://www.dlr.de/dlr/Portaldata/1/Resources/bilder/portal/portal_2012_1/leitstudie2011_bf.pdf, access 01 June 2017

32. Gender Pay Gap in the Highest Salary Group in the Energy Sector

Justification referring to Sustainability Concept	This indicator addresses the sustainability rule *'Equal access for all to information, education, and occupation'* for the energy system and the energy transition: Granting equal opportunities primarily relates to equality between gender, ethnicity, and cultural background, mainly in the field of employment and income conditions in the energy sector.
Description	The gender pay gap is presented in terms of gross yearly income of fulltime employees in the energy supply sector including special payments (acc. to the German Federal Statistical Office category 'D-Energy supply' including electricity, gas, heat, cold supply).
Unit	€/a
Data	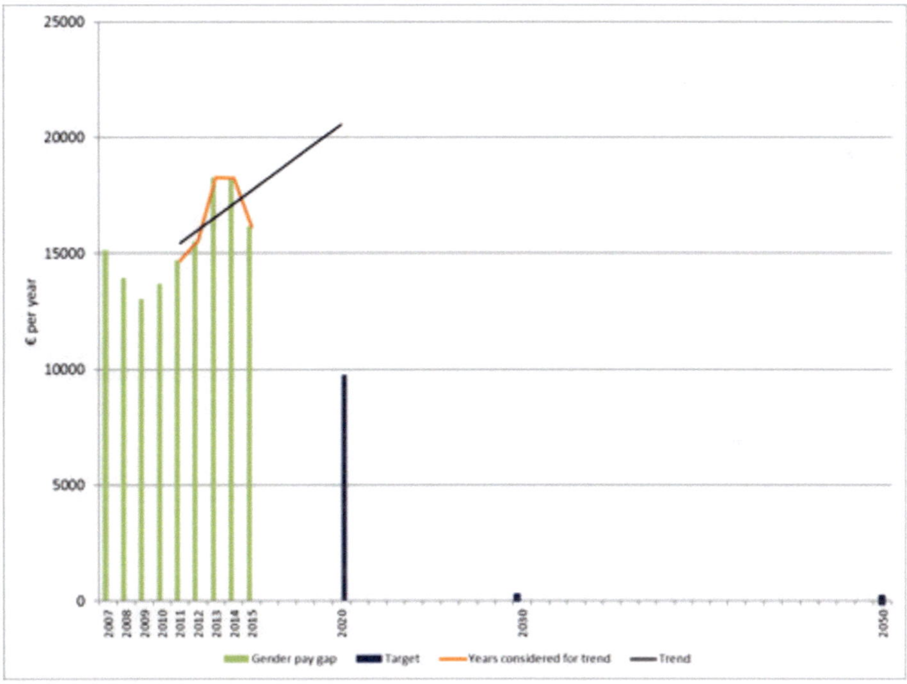 Figure 1: Gender pay gap in €/a for 2007 to 2015 (own diagram based on [1-2])
Targets	2020: 9,754 €/a 2030: 0 €/a 2050: 0 €/a
Assessment	 Latest available value (2015): 16,172 € in gross yearly salary [1] The trend calculated based on the previous 5 years (2011–2015) shows an increase in the gender pay gap of 24 % by the year 2020 compared to the reference value (the average value over the years 2011 to 2015). The target value, however, requires a decrease to '0' by the year 2030. Thus, the indicator is assessed with a red traffic light, meaning there is a high probability that the target of 2020 will be missed.

Comments Data	In 2015, women's salary amounted to 84 % of men's salary with an annual salary differential of around 16,000 Euros. Official statistics distinguish between five performance groups. The performance groups defined represent a rough categorization of the employees' activities according to the qualification profile of the workplace. This was narrowed down to the 'highest salary group' for clearer visualization and relevance and to ensure continuous supply of reliable data from the Federal Statistical Office. Performance group 1 includes 'employees in a leading position' (employees with supervisory and discretionary authority) such as employed managers, provided their earnings include, at least partially, non-performance-related payments. It also includes employees in larger management areas who perform dispatching or management tasks as well as employees with activities that require comprehensive business or technical expertise. In general, the specialist knowledge is acquired through university studies.
Comments Targets	Until 2030, the gender pay gap should be '0'. The value for 2020 results from interpolation between the average value of the previous five years (2011–2015) and the value for 2030 ('0').
Literature	[1] Destatis: Statistisches Bundesamt: Verdienste und Arbeitskosten – Arbeitnehmerverdienste. Fachserie 16 Serie 2.3, 2015. https://www.destatis.de/DE/Publikationen/Thematisch/VerdiensteArbeitskosten/Arbeitnehmerverdienste/ArbeitnehmerverdiensteJ2160230157004.pdf?__blob=publicationFile, access 8 March 2017 [2] Destatis: Statistisches Bundesamt: Fachserie. 16, Verdienste und Arbeitskosten. 2. 1. Arbeitnehmerverdienste 2007 – 2016, https://www.destatis.de/GPStatistik/receive/DESerie_serie_00000297?list=all, access 15 March 2017

33. Share of Regulatory Tools in the Planning of Power Transmission Grids Which Fulfill Regulatory Requirements

<table>
<tr>
<td>Justification referring to Sustainability Concept</td>
<td>This indicator addresses the sustainability rule 'Participation in societal decision-making processes': All members of society should be able to participate in decision processes relevant to energy issues, particularly referring to planning processes. This requires transparency, openness of procedures and access to suitable information, as well as enabling institutional structures for deliberation and participation formats tailored to energy issues as far as necessary.</td>
</tr>
<tr>
<td>Description</td>
<td>We propose a monitoring of objective minimum criteria in planning and approval procedures for energy infrastructure projects. The following requirements should be part of the monitoring: firstly, public participation/integration; secondly, availability of relevant documents to the population, which, among other things, should definitely include a report on the project's likely impact on the environment, the social situation and public finances, i.e., a sustainability report [1]; thirdly, a public hearing during which objections and critiques are debated and negotiated; fourthly, involvement of a mediator in the decision-making process, and fifthly, the right to contest a decision.

These five criteria should be met in the following steps of the energy planning:

(1) Definition of the objectives and principles of energy policy
(2) Preparation of the energy demand plan for Germany
(3) Preparation of the German network development plan
(4) Planning and approval process for energy projects
(5) Preliminary environmental impact assessment (EIA)

Due to the importance of the electricity grid planning, we suggest to monitor this planning process.</td>
</tr>
<tr>
<td>Unit</td>
<td>%</td>
</tr>
<tr>
<td>Data</td>
<td>

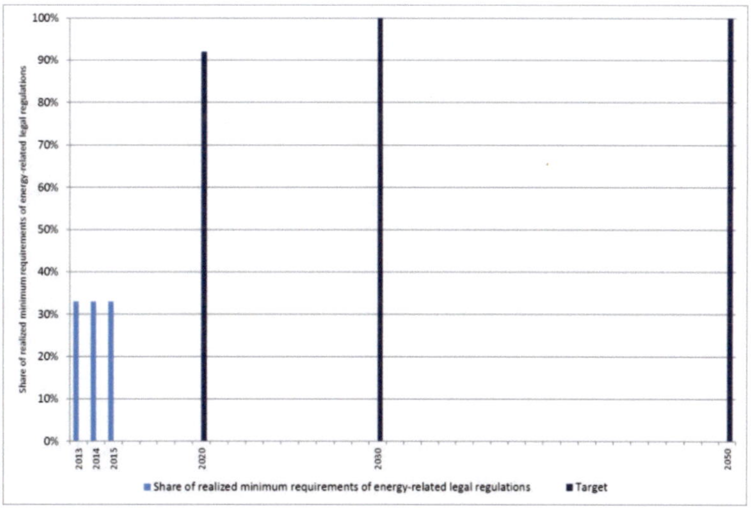

Figure 1: Fulfillment of minimum requirements of legal regulations in power grid planning (own diagram based on [2-3])
</td>
</tr>
</table>

See Table 1 for further information.

Table 1: Fulfillment of minimum requirements of legal regulations in power grid planning in 2015

Energy planning stage	**Public participation**	**Complete documentation for the population**	**Sustainability studies in the assessment of alternatives**	**Public event**	**Mediator**	**Right to contest**
Objectives and principles of the Government's energy policy	(r)1	r	–	r	–	–
Federal requirements plan	e	e	r	r	r	r
Network development plan	e	e	(r)2	r	r	r
Planning and approval process	e	e	–	e	e	
Preliminary EIA	(r)2	r	–	r	r	r

r = required, e = existent, – = not necessary

1 = No public participation, but drafts are published.
2 = Only an environmental impact study is carried out. An evaluation of other sustainability aspects is not provided for.

Targets

2020: 92 %
2030: 100 %
2050: 100 %

Assessment

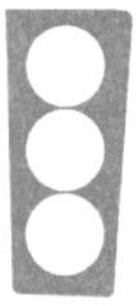

Latest available value (2015): 33 % (own calculation after [2] and [3])

In 2015, 8 out of 24, i.e., 33 % of the characteristics of good participation opportunities were met. Due to the lack of data series, an assessment is not possible

Comments Data

The expectations towards public participation are high: Involving the public into decision making is supposed to improve transparency and quality of the decision-making process and to generate legitimation for the decision. Ultimately, this should result in a higher acceptance of energy infrastructure projects [4]. In addition, this process can identify projects facing grave problems of acceptance so that it might be better to change or stop these project rather than investing more money and effort in them.

This indicator focuses on the evaluation of objective minimum requirements in decision-making processes in energy infrastructure projects. The influence on decision making depends on the form of participation: from public information and hearings to the power of veto. Also, the public can be involved at different project stages: from requirements and design planning to the approval procedure. Finally, the results of public participation are not necessarily binding [5].

	Key to the energy projects is the Federal Government's energy program, since it includes specifications referred to by other plans such as the federal requirements plan, the act on energy demand planning or the network development plan [6]. Al-though the Federal Government in the representative democracy of Germany is free to determine the principles and guidelines of energy policy, it would be desirable from points of view of participation to involve the public in determining these principles and guidelines. This should be accompanied by an information process, including a public event and the provision of relevant documents for the decision. However, this process need not necessarily involve a mediator, and the decision also does not need to be contestable, if only because the program has no legally binding effect. The situation is different in energy demand planning and network development planning. In Germany, the planning of power grids involves a four-step process [3]. Annually, scenarios for the development of the German energy industry are prepared and, based on that, network development plans are developed. In these two steps, the public is repeatedly involved. In the last round of participation, estimates of the environmental impact are presented for one of the scenarios. Thus, participation opportunities are already quite good in these two process steps. However, the process could be further improved by conducting a public hearing involving a mediator and by preparing an EIA also for those scenarios that are not intended to be pursued. In addition, the demand planning decision should be contestable in order to check as early as possible whether the result was influenced by errors made in assessing and determining the demand. The network development plan is presented to the Federal Government at least every three years. The Government deliberates on it and adopts the federal demand plan [7]. This is legally binding for the subsequent planning. After becoming effective, this document will be developed further with public participation. For concrete energy projects such as the specific planning of a power line, zoning and planning approval processes are undertaken where the Environmental Impact Assessment Act (Gesetz über die Umweltverträglichkeitsprüfung – UVPG) applies and where the opportunities for public participation are relatively well regulated and institutionalized. Here, the only need for improvement is seen in switching over from an environmental impact assessment to a sustainability impact assessment. EIA legislation, in principal, also applies to the planning of a wind farm or a biogas plant. This legislation provides that an environmental impact assessment must be carried out for plants of a certain size. It also generally requires that the public is informed and involved in the course of the EIA. However, many energy projects are excluded from a general requirement for EIA, such as wind farms under 20 wind turbines, biogas plants, geothermal systems, and marine energy parks [2]. For some of these plants, a preliminary examination must be carried out to decide on whether an EIA is required or not (§ 3c Abs. 2 UVPG). It is envisaged that the public will be informed of this preliminary examination and any negation of the EIA requirement. Usually, it is not envisaged to involve the public in the preliminary examination. However, public participation also in this process would be desirable. For this purpose, the documents used in decision making should be made public, there should be an event for public participation, and the decision should be contestable. In principle, it should be possible to involve a mediator. If in the future similar arrangements are planned for gas networks and when setting up the 'National plan for competitive, secure, and sustainable energy' for Germany, as planned by the EU [8], the above-mentioned criteria should be applied.

Comments Targets	To fulfill the minimum requirements, it is necessary to implement the measures listed in the Federal Requirements Plan Act and subsequent laws. The implementation of the sustainability report is an exception. Further research and implementation work is needed on this, inter alia, to determine what information should be provided in such a report. Its specification and translation into legal requirements could be completed by 2030. For this reason, it is a requirement of the monitoring to implement all elements presented by 2020 and to have the sustainability impact assessment implemented by 2030.
Literature	[1] Grunwald, A.; Kopfmüller, J.: Die Nachhaltigkeitsprüfung: Kernelemente einer angemessenen Umsetzung des Nachhaltigkeitsleitbilds in Politik und Recht. Wissenschaftliche Berichte FZKA 7349, Karlsruhe, 2007 [2] Bundesministerium der Justiz und für Verbraucherschutz: Gesetz über die Umweltverträglichkeitsprüfung (UVPG). Sachstand 20.11.2015, Berlin, 2015 [3] 50Hertz Transmission GmbH; Tennet TSO GmbH; Amprion GmbH; TransnetBW GmbH: Konsultationsleitfaden. Berlin, 2015 [4] Schweizer, P.-J.; Renn, O.: Partizipation in Technikkontroversen: Panakeia für die Energiewende? In: Technikfolgenabschätzung – Theorie und Praxis 22(2), 2013, p. 43 [5] Kompetenzzentrum Öffentliche Wirtschaft, Infrastruktur und Daseinsvorsorge e.V.: Optionen moderner Bürgerbeteiligung bei Infrastrukturprojekten – Ableitungen für eine verbesserte Beteiligung auf Basis von Erfahrungen und Einstellungen von Bürgern, Kommunen und Unternehmen. 2013, p. 54. http://www.wifa.uni-leipzig.de/fileadmin/user_upload/KOZE/Downloads/Optionen_moderner_Bu%CC%88rgerbeteiligungen_bei_Infrastrukturprojekten_.pdf, access 10 February 2015 [6] Hermes, G.: Planungsrechtliche Sicherung einer Energiebedarfsplanung – ein Reformvorschlag. ZUR (5/2014), 2014, p. 259–269 [7] Bundesministerium der Justiz und für Verbraucherschutz: Gesetz über den Bundesbedarfsplan. Berlin, 2013 [8] EC: EU 2030 strategy. 2015. https://ec.europa.eu/energy/en/topics/energy-strategy/2030-energy-strategy, access 3 April 2016

34. Share of Tourists Who Perceive Energy Power Technologies as Disruptive in the Vacation Area

<table>
<tr><td>Justification referring to Sustainability Concept</td><td>This indicator addresses the sustainability rule 'Conservation of the cultural function of nature': Cultural and natural landscapes in general and landscape elements with particular characteristic features and beauty are to be conserved. Energy infrastructures can impair recreational values, spiritual and sensual meanings or aesthetic contemplation potentials of nature.</td></tr>
<tr><td>Description</td><td>The data are derived from an annual survey, the 'Reiseanalyse RA 2014'. The results are representative for leisure travelers within the German-speaking residential population (N=54.8 million, n=6,070). The questions relate to the destination of the main holiday trip in 2013. For the first time, the negative perception of visual pollution by energy infrastructure was included in this survey [1].</td></tr>
<tr><td>Unit</td><td>%</td></tr>
<tr><td>Data</td><td>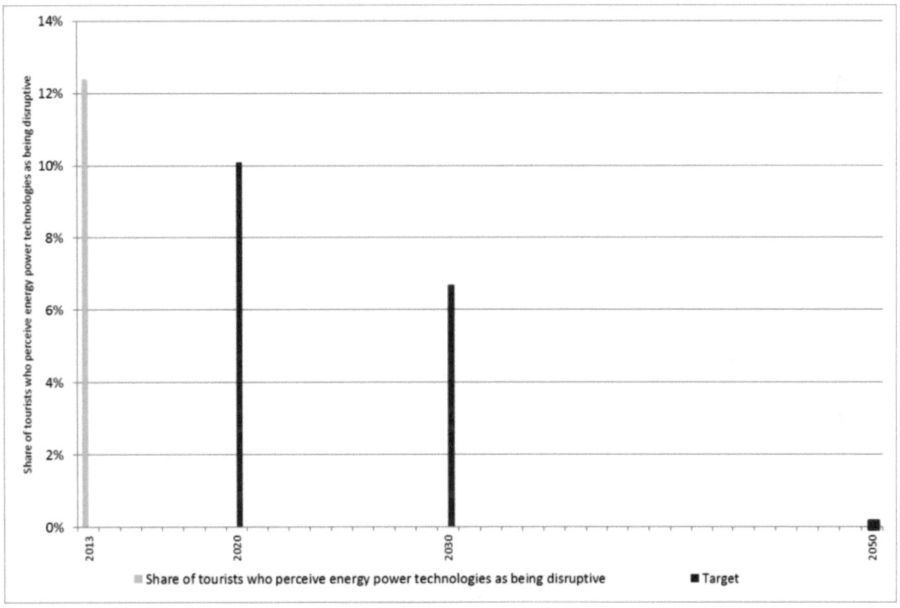

Figure 1: Share of tourists who perceive energy power technologies as disruptive in the vacation area (own diagram based on [2])</td></tr>
<tr><td>Targets</td><td>2020: 10 %
2030: 7 %
2050: 0 %</td></tr>
<tr><td>Assessment</td><td>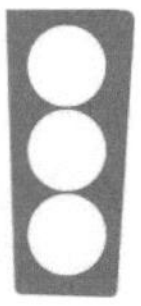
Latest available value (2013): 12.4 % [2]
There is no assessment possible due to a lack of data.</td></tr>
</table>

Comments Data	The visual impact of energy infrastructure and, accordingly, the data collection was differentiated into three groups: Renewable Energies (RE), RE/Non-RE and the group of Non-RE [1]. The corresponding items included in the RE group are: separate wind turbines, wind farms, solar parks, roof-top solar plants, and biogas plants. Within the mixed group RE/Non-RE, monotonous agricultural land use and power lines are considered. The group of Non-RE is composed of industry chimneys and cooling towers, high-rise buildings, and motor and express ways. Looking in more detail at the three different groups mentioned above shows that 8.3 % of the tourists perceived elements of the RE group as disruptive, 3.2 % were disturbed by elements of the mixed RE/Non-RE group, and 4.8 % by the Non-RE group, respectively [3]. Within RE, wind power installations were perceived as disruptive by 7.1 %, solar power infrastructure by 2.0 %, and biogas plants by 1.8 % of the tourists [3]. It was possible to give more than one answer. The collection of data on tourism, renewable energies and landscape will also be part of the 'Reiseanalyse 2015'. Therefore, the items have been adjusted in some parts: Wind farms will be differentiated into on- and offshore, and the three items within the Non-RE group will no longer be part of the evaluation. They are replaced by the item 'power plants (coal and nuclear).' Thus, conventional energy production infrastructure will be part of the evaluation although this makes updating the data more difficult.
Comments Targets	In order to achieve the energy transition, it is important that the measures to be implemented are accepted. The acceptance by tourists is of particular importance, especially in areas where tourism makes an important contribution to the economic output, since it is obvious that if the measures are not accepted by the tourists, resistance against the measures will also develop among the population depending on tourism. For this reason, the possible acceptance of the project by tourists should be considered in the choice of location, the dimensioning and design of energy technologies – especially in tourist regions. Currently (2013), the share of tourists who perceive energy power technologies as disruptive in the vacation area is 12.4 %. This value should decrease to 'zero' by 2050. The values for the years 2020 and 2030 were interpolated accordingly.
Literature	[1] NIT Institut für Tourismus- und Bäderforschung in Nordeuropa GmbH: Erneuerbare Energien und Tourismus in Mecklenburg-Vorpommern, 2014. http://www.nit-kiel.de/fileadmin/user_upload/Grimm-Handout-EE-MV-20140707.pdf, access 3 April 2016 [2] NIT Institut für Tourismus- und Bäderforschung in Nordeuropa GmbH: Tourismus, Erneuerbare Energien und Landschaftsbild. Eine Studie auf Basis der Reiseanalyse RA 2014. Kiel, 2014 [3] NIT Institut für Tourismus- und Bäderforschung in Nordeuropa GmbH: Einflussanalyse Erneuerbare Energien und Tourismus in Schleswig-Holstein – Kurzfassung der Ergebnisse. 2014. http://www.wind-energie.de/sites/default/files/attachments/region/schleswig-holstein/20140722-ee-tourismus-sh-kurzfassung.pdf, access 3 April 2016

35. Acceptance of Renewable Energies in the Neighborhood

<table>
<tr><td>Justification referring to Sustainability Concept</td><td>This indicator addresses the sustainability rule 'Conservation of social resources': To ensure the social cohesion of society with respect to the energy system, an 'environment' of lawfulness and justice, tolerance, solidarity, and common welfare orientation should be provided, and potentials of non-violent conflict resolution should be strengthened. Well-organized discourses represent means to achieve rational agreements and facilitate fair societal settlement of interests in energy-related decisions.</td></tr>
<tr><td>Description</td><td>Based on different surveys in several years, the acceptance of various aspects of the energy system was analyzed on behalf of the Renewable Energies Agency (Agentur für Erneuerbare Energien) [1]. Data are available for Germany and for each of the 16 federal states for the years 2010, 2011, 2012 and for the years 2013 to 2016 [2, 3, 4, 5].</td></tr>
<tr><td>Unit</td><td>%</td></tr>
<tr><td>Data</td><td>
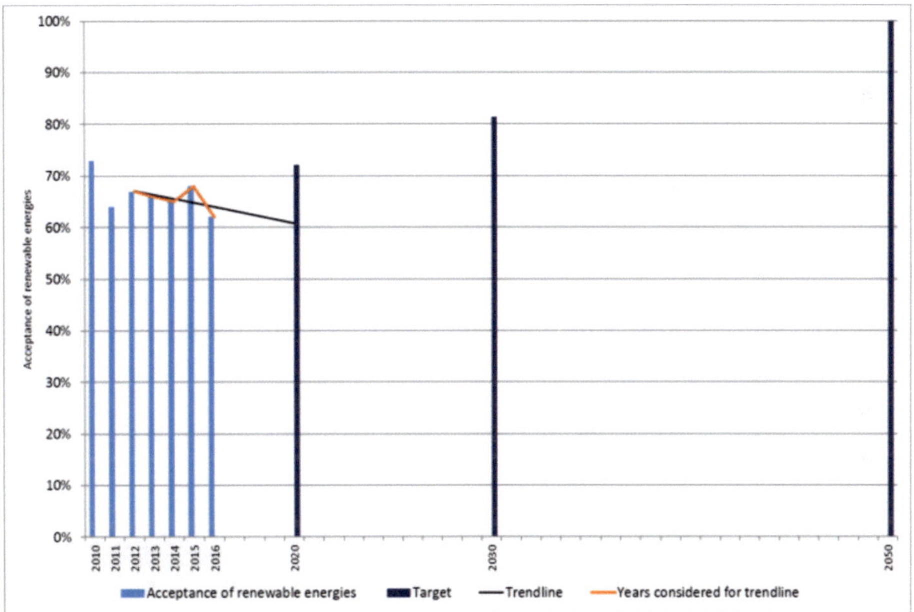

Figure 1: Acceptance of renewable energy in the neighborhood in general (own diagram based on [1-5])</td></tr>
<tr><td>Targets</td><td>2020: 71 %
2030: 81 %
2050: 100 %</td></tr>
<tr><td>Assessment</td><td>
Latest available value (2016): 62 % [2]
The trend calculated based on the previous 5 years (2012–2016) shows a decrease in the acceptance of renewable energy in the neighborhood of 7.3 % by 2020 compared to the reference value (the average value over the years 2012 to 2016). The target value for 2020 requires an increase of 8.7 % compared to the reference value. A red traffic light is given, since the expected trend goes in the wrong direction not reaching the target of 2020.</td></tr>
</table>

Comments Data

In addition to the acceptance of renewable energy in the neighborhood, data are also available for the acceptance of nuclear power plants, coal-fired power plants, renewable energy in general, wind turbines, biomass, and photovoltaic systems (solar parks) as well as for the acceptance of grid extension (see Indicator 36). The percentages given are the sum of the answers: 'I like that' and 'I like that very much'. For a comparison, these data are shown in Table 1.

Table 1: Acceptance of different technologies in the neighborhood

	Acceptance in the neighborhood (%)						
	2010	2011	2012	2013	2014	2015	2016
Solar park	74	77	77	72	72	77	73
Biomass	40	36	36	39	39	39	38
Nuclear power plant	5	2	3	3	5	4	5
Coal-fired power plant	6	8	8	8	11	7	6
Wind turbines	56	60	61	59	61	59	52

Data taken from [1-5]

Comments Targets

A linear interpolation to 100 % in the year 2050 was chosen, starting from the average value over the previous 5 years (2012–2016).

Literature

[1] Agentur für Erneuerbare Energien: Bundesländer in der Übersicht. https://www.foederal-erneuerbar.de/uebersicht/bundeslaender/BW|BY|B|%20BB|HB|HH|HE|MV|NI|NRW|RLP|SL|SN|ST|SH|TH|D/kategorie/akzeptanz/auswahl/229-gesellschaftliche_ak/#goto_229, access 7 May 2015

[2] Agentur für Erneuerbare Energien: Umfrage 2016: Bürger befürworten Energiewende und sind bereit, die Kosten dafür zu tragen. https://www.unendlich-viel-energie.de/mediathek/grafiken/akzeptanz-umfrage-2016, access 8 March 2017

[3] Agentur für Erneuerbare Energien: Umfrage 2013: Bürger befürworten Energiewende und sind bereit, die Kosten dafür zu tragen. http://www.unendlich-viel-energie.de/themen/akzeptanz2/akzeptanz-umfrage/umfrage-2013-buerger-befuerworten-energiewende-und-sind-bereit-die-kosten-dafuer-zu-tragen, access 7 June 2016

[4] Agentur für Erneuerbare Energien: Umfrage Akzeptanz Erneuerbarer Energien 2014. http://www.unendlich-viel-energie.de/mediathek/grafiken/akzeptanzumfrage-erneuerbare-energie-2014, access 7 June 2016

[5] Agentur für Erneuerbare Energien: Umfrage zur Akzeptanz Erneuerbarer Energien 2015. http://www.unendlich-viel-energie.de/mediathek/grafiken/umfrage-akzeptanz-erneuerbare-energien-2015, access 7 June 2016

36. Acceptance of Grid Extension for Achieving 100 % Renewable Energy Supply

Justification referring to Sustainability Concept	This indicator addresses the sustainability rule *'Conservation of social resources'*: To ensure the social cohesion of society with respect to the energy system, an 'environment' of lawfulness and justice, tolerance, solidarity, and common welfare orientation should be provided, and potentials of non-violent conflict resolution should be strengthened. Well-organized discourses represent means to achieve rational agreements and facilitate fair societal settlement of interests in energy-related decisions.
Description	With the increase of electricity generated from RE, increasing amounts of renewable energy must be integrated into the power grid and – if distance increases between production and consumption sites – transported over long distances. This will only be achieved by targeted expansion of transmission and distribution networks. However, this often meets with resistance from the population concerned.
Unit	%
Data	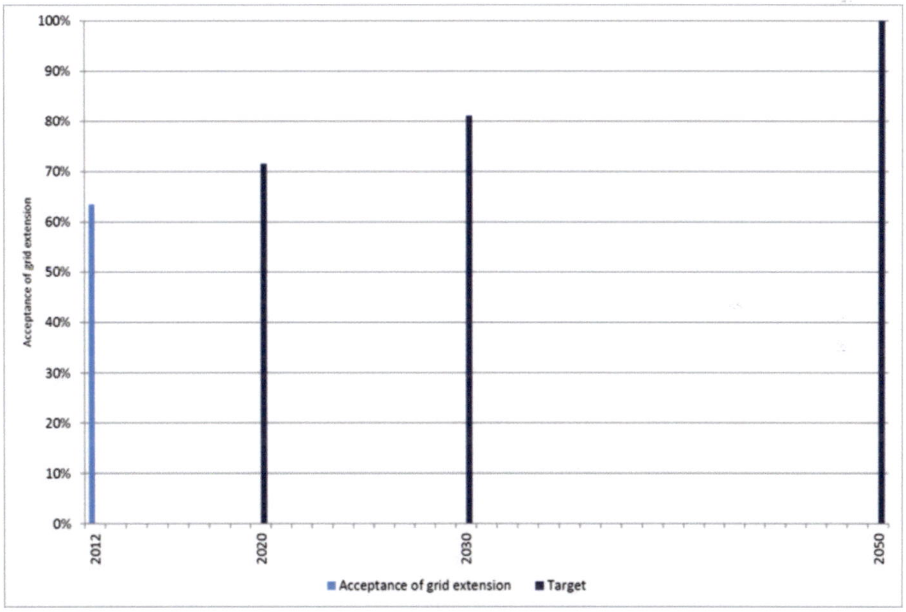 Figure 1: Acceptance of grid extension for achieving 100 % renewable energy supply (own diagram based on [1])
Targets	2020: 72 % 2030: 81 % 2050: 100 %
Assessment	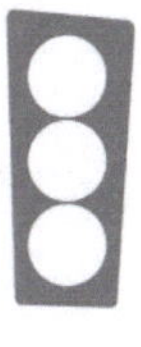 Latest available value (2012): 63.5 % [1] There is no assessment possible because data are available only for the year 2012.

Comments Data	Data on grid extension for Germany based on a survey are available only for the year 2012. Data are also available in [1] for: • Grid extension as a precondition for transportation of regionally produced renewable energy: The acceptance in 2012 was 62.7 % in Germany. • Grid extension via underground cable: Acceptance in 2012 was 76.6 % in Germany.
Comments Targets	A linear interpolation to 100 % in the year 2050 was chosen, starting from the value for the year 2012.
Literature	[1] Agentur für Erneuerbare Energien: Bundesländer in der Übersicht. Bundesländer-Übersicht zu Erneuerbaren Energien. https://www.foederal-erneuerbar.de/uebersicht/bundeslaender/BW|BY|B|%20BB|HB|HH|HE|MV|NI|NRW|RLP|SL|SN|ST|SH|TH|D/kategorie/akzeptanz/auswahl/504-akzeptanz_von_netzau/#goto_504, access 7 May 2015

37. Degree of Internalization of Energy-related External Costs

Justification referring to Sustainability Concept	This indicator describes the instrumental rule *'Internalization of external social and ecological costs'*: Implementing the 'polluter pays' principle, energy-related prices and regulations must reflect external environmental and social costs caused by the energy system. To evaluate the degree of internalization, information is needed about the amount of these costs and mechanisms of internalization (e.g., via taxes).
Description	The share of taxes on energy use (energy taxes, electricity taxes, motor vehicle taxes, air transport taxes, nuclear fuel taxes, road taxes) in environmental costs due to electricity production and energy use for heating and transportation is defined as the degree of internalization of energy-related external costs. Data are given for the years 2008 to 2010 and are calculated from information given in [1].
Unit	%
Data	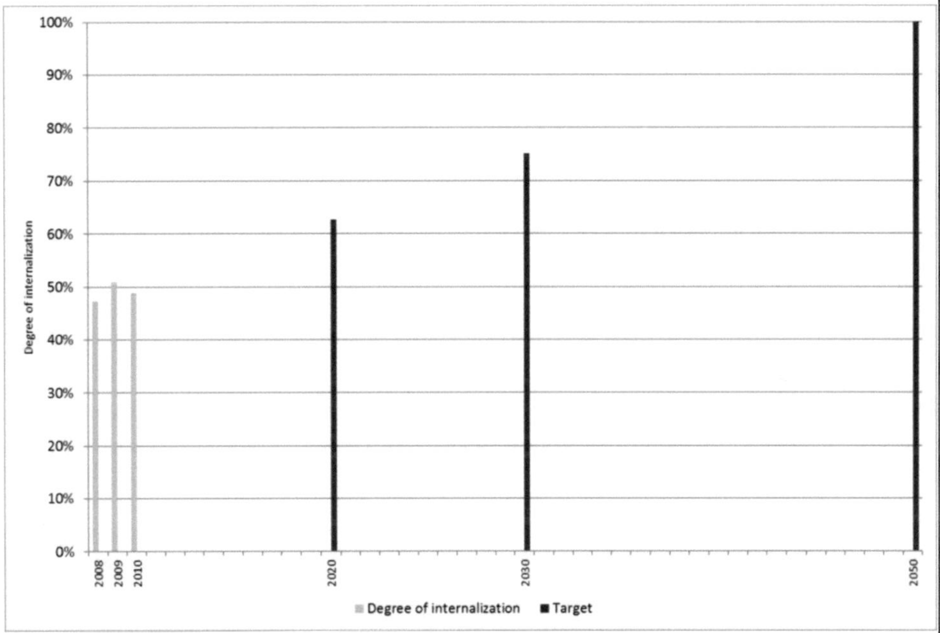 Figure 1: Internalization of energy-related external costs (own diagram based on [1-3])
Targets	2020: 63 % 2030: 75 % 2050: 100 %
Assessment	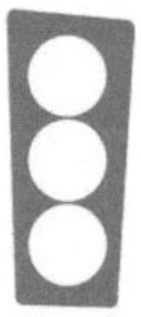 Latest available value (2010): 48.9 % [1-3] Data are only available for the years 2008 to 2010. Based on these 3 years, no trend can be calculated. However, the degree of internalization should increase, which will be possible if additional taxes are introduced or if emissions are reduced.

Comments Data	Data on energy taxes, electricity taxes and motor vehicle taxes are taken from [2,3], data on road taxes for trucks are taken from [4]. Data on environmental costs are taken from [1]. According to [5], environmental costs resulting from the production of electricity in Germany include environmental costs as well as costs related to human health that result from direct emissions. Costs resulting from indirect emissions over the entire life cycle have also been taken into consideration. Since indirect emissions arise not only in Germany, EU cost rates have been considered as well. The costs of GHG emissions are 80 € per t CO_2 (including damage as well as abatement costs). Estimates of environmental costs of nuclear energy differ widely in the available literature. Following the requirements of the Methodological Convention (Methodenkonvention) [5], the most expensive technology should be used for the calculations. In the case considered here, this is lignite. Environmental costs of transportation include health effects, climate change effects, noise, and impact on nature and landscape as well as effects due to indirect emissions (construction, maintenance and disposal, fuel supply). Environmental costs amount to 122.4 billion € in 2008, 115.2 billion € in 2009, and 120.6 billion € in 2010 [1]. In general, data for other years can also be calculated by taking into consideration the mix of electricity production, heat energy consumption, as well as the relevant data for the transport sector for the different years. However, this is only reasonable if both the related environmental costs and the technologies (e.g., emission factors) do not change.
Comments Targets	All external costs should be internalized. Therefore, in 2050 this value should be 100 %. The values for the years 2020 and 2030 are the result of a linear interpolation between the average value over the years 2008 to 2010 and the value for the year 2050.
Literature	[1] UBA – Umweltbundesamt: Daten zur Umwelt 2015, August 2015. https://www.umweltbundesamt.de/sites/default/files/medien/376/publikationen/daten_zur_umwelt_2015.pdf, access 01 June 2017 [2] Bundesfinanzministerium: Kassenmäßige Steuereinnahmen nach Steuerarten in den Kalenderjahren 2006 – 2009, http://www.bundesfinanzministerium.de/Content/DE/Standardartikel/Themen/Steuern/Steuerschaetzungen_und_Steuereinnahmen/2012-05-29-steuereinnahmen-nach-steuerarten-2006-2009.pdf?__blob=publicationFile&v=3, access 22 March 2017 [3] Bundesfinanzministerium: Kassenmäßige Steuereinnahmen nach Steuerarten in den Kalenderjahren 2010 – 2016 http://www.bundesfinanzministerium.de/Content/DE/Standardartikel/Themen/Steuern/Steuerschaetzungen_und_Steuereinnahmen/2017-05-05-steuereinnahmen-nach-steuerarten-2010-2016.pdf?__blob=publicationFile&v=5, access 01 June 2017 [4] Statista: Mauteinnahmen in Deutschland von 2005 bis 2016* (in Milliarden Euro). http://de.statista.com/statistik/daten/studie/75600/umfrage/mauteinnahmen-in-deutschland-seit-2005 /, access 23 December 2015 [5] UBA – Umweltbundesamt: Ökonomische Bewertung von Umweltschäden – Methodenkonvention 2.0 zur Schätzung von Umweltkosten, August 2012. https://www.umweltbundesamt.de/sites/default/files/medien/378/publikationen/uba_methodenkonvention_2.0_-_2012_gesamt.pdf, access 01 June 2017

38. Share of Development Aid Expenses for Energy-related Projects in Total GDP

Justification referring to Sustainability Concept	This indicator addresses the instrumental rule *'Promotion of international co-operation'*: To implement the required global energy system transitions, the different actors (governments, companies, NGOs) need to work together in a spirit of global partnership, including the creation of necessary political, legal, and factual framework conditions. This also addresses the question of how to define and implement globally differentiated responsibilities to cope with energy-related challenges such as greenhouse gas emissions, lacking access to affordable energy services, etc.
Description	Flows of official financing administered to promote the economic development and welfare of developing countries in relation to the GDP.
Unit	%
Data	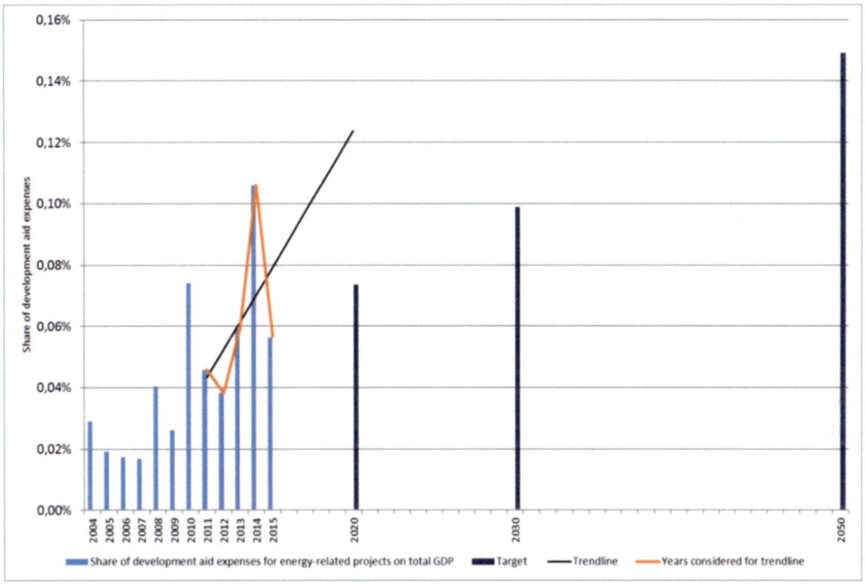 Figure 1: Share of development aid expenses for energy-related projects in total GDP for Germany (own diagram based on [1])
Targets	2020: 0.07 % 2030: 0.09 % 2050: 0.15 %
Assessment	 Latest available value (2015): 0.06 % [1] The trend calculated based on the previous 5 years (2011–2015) shows an increase of the share for energy related projects in total GDP of about 102 % until 2020 compared to the reference value (the average value over the years 2011 to 2015). The target value for 2020 requires an increase of only 20 % compared to the reference value. Due to the expected overachievement of the target of 2020, the indicator is assessed with a green traffig light, meaning that the achievement of the target of 2020 is likely.

Comments Data	ODA is defined in the following way: 'Flows of official financing administered with the promotion of the economic development and welfare of developing countries as the main objective, and which are concessional in character with a grant element of at least 25 percent (using a fixed 10 percent rate of discount). By convention, ODA flows comprise contributions of donor government agencies, at all levels, to developing countries ('bilateral ODA') and to multilateral institutions. ODA receipts comprise disbursements by bilateral donors and multilateral institutions' [2]. Therefore, ODA needs to fulfill these three requirements [3] • undertaken by the official sector (official agencies, including state and local governments, or their executive agencies), • promotion of economic development and welfare as the main objective, • at concessional financial terms (if a loan, having a grant element of at least 25 percent). From the ODA expenditures of Germany, the expenditures for the energy sector (sector II.3 – energy in the official data base) are selected. This number is compared to the GDP. As data for ODA are given in US$, data for the GDP given in US$ in PPP (purchasing power parity) are used. Data for ODA (total ODA and ODA for the energy sector) are taken from official OECD statistics [1]. Data for GDP in PPP are taken from [4–5].
Comments Targets	Over the previous 5 years, the average of the share of ODA in the GDP was 0.47 % (0.48 % in 2014 and 0.45 % in 2015). According to the UNEP's Sustainable Development Goals, developed countries should 'implement fully their ODA commitments, including to provide 0.7 % of GNI in ODA to developing countries of which 0.15–0.20 % to least-developed countries' [6]. Also the German Advisory Council on Global Change (WBGU) recommends in its 2003 assessment that 'the federal government should considerably increase its ODA funds beyond the 0.33 % which were announced until 2006' and suggests 'spending at least 0.5 % of the GDP on ODA until 2010. According to the current pressure, even an increase to approx. 1 % of the GDP would be appropriate' ([7], p. 188). The average share of energy-related ODA projects in the GDP was 0,061 % over the previous 5 years (2011–2015). Assuming an increase of the overall ODA share in the GDP should increase to the stated target value of 1 % by 2050 and the share of energy-related projects remains the same (0.061 %), this results in the following target values for the 'share of ODA for energy-related projects'.
Literature	[1] OECD.StatExtracts: Aid (ODA) by sector and donor [DAC5]. http://stats.oecd.org/Index.aspx?datasetcode=TABLE5, access 7 February 2017 [2] OECD: Official Development Assistance (ODA). http://stats.oecd.org/glossary/detail.asp?ID=6043, access 7 February 2017 [3] Wikipedia: Official development assistance. http://en.wikipedia.org/wiki/Official_development_assistance#cite_note-2, access 13 May 2015 [4] The World Bank: GDP per capita, PPP (current international $). http://data.worldbank.org/indicator/NY.GDP.PCAP.PP.CD?page=1, access 7 February 2017

	[5] The World Bank: population, total. http://data.worldbank.org/indicator/SP.POP.TOTL, access 7 February 2017 [6] UNEP Introduction to the Proposal of The Open Working Group for Sustainable Development Goals, http://geodata.grid.unep.ch/gegslive/sdgtext.php, access 30 Nov 2016 [7] WBGU – Wissenschaftlicher Beirat der Bundesregierung Globale Umweltveränderungen: Welt im Wandel. Energiewende zur Nachhaltigkeit. Berlin-Heidelberg, 2003, p. 188

39. Share of Households Producing Renewable Electricity

Justification referring to Sustainability Concept	This indicator addresses the sustainability rule *'Society's ability to respond'*: The ability of societies to identify and suitably react to problems in environmental and social systems arising from the energy system and to promote the energy transition has to be maintained or improved. This requires answering questions of relevance, particularly for which developments or events society should be prepared, and establishing suitable information, communication, institutional and organizational structures.
Description	Share of households producing renewable electricity
Unit	%
Data	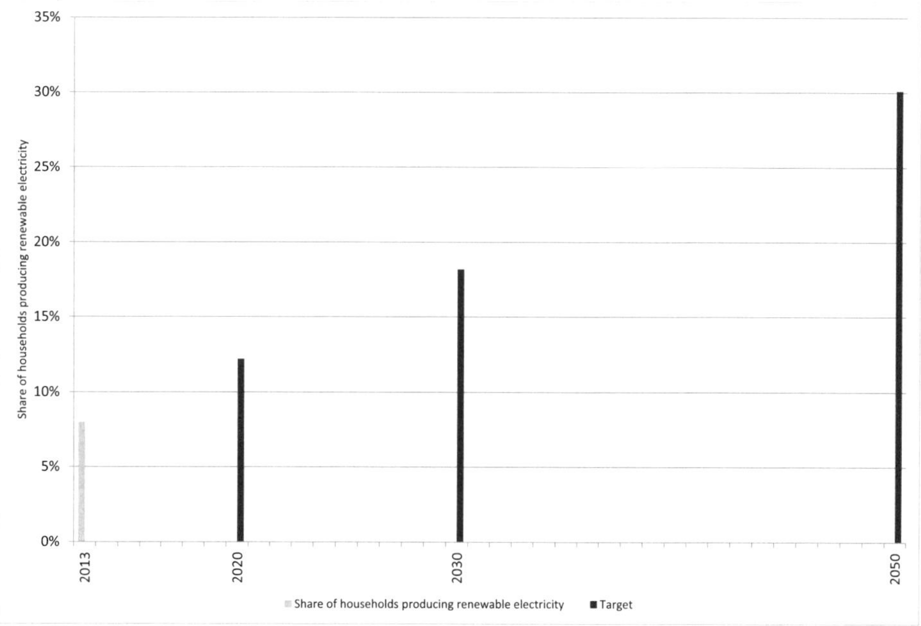 Figure 1: Share of households producing renewable electricity for their own use (own diagram based on [1])
Targets	2020: 12 % 2030: 18 % 2050: 30 %
Assessment	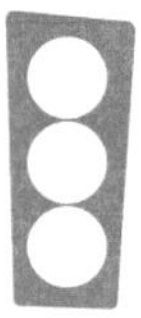 Latest available value (2013): 8 % [1] There is no assessment possible because only one data record is available.
Comments Data	Data base on a representative survey conducted by TNS Emnid on behalf of E WIE EINFACH. Only data for 2013 are given.

Comments Targets	Justification: If a share of 80 % of renewables shall be reached in 2050, it is necessary to use all those available possibilities to produce renewable energy which have no significant negative side effects. One of these resources is harvesting sun energy from roofs and walls of new and existing buildings. This option has only little negative influence on other sustainability rules. Nevertheless, the installation of PV on roofs and walls and the use of combines heat and power facilities, together with local storage systems, might contribute to the downsizing of electricity infrastructures like grid or regional storage systems and make houses and neighborhoods more resilient to blackouts. However, only people who own their residence are able to install devices for energy production at or on top of the building. Approx. 43 % of all German households are living in their own property (2013, [2]). It also has to be kept in mind that not every building is suitable for energy production due to shadowing, the protection of historic buildings, or for other reasons. Against this background we pursue the goal to use 70 % of the owner-occupied buildings for the production of energy by 2050. Since the available data only refer to the share in all households, the target value proposed for 2050 is 70 % of 43 %. These are 30 %. The values for 2020 and 2030 were calculated as linear interpolation between the 2050 target value and the value for 2013.
Literature	[1] Agentur für Erneuerbare Energien, Föderal Erneuerbar: Anteil der Haushalte mit eigenproduziertem Strom. 2013. https://www.foederal-erneuerbar.de/landesinfo/bundesland/D/kategorie/akzeptanz/auswahl/655-anteil_der_haushalte/#goto_655, access 3 April 2016 [2] Destatis: Wirtschaftsrechnungen. Einkommens- und Verbrauchsstichprobe Wohnverhältnisse privater Haushalte. Fachserie 15, Sonderheft 1, 2013, p. 20. https://www.destatis.de/DE/Publikationen/Thematisch/EinkommenKonsumLebensbedingungen/Wohnen/EVS_HausGrundbesitzWohnverhaeltnisHaushalte2152591139004.pdf?__blob=publicationFile, access 01 June 2017

40. Share of Households Buying Renewable Electricity

<table>
<tr><td>Justification referring to Sustainability Concept</td><td>This indicator addresses the instrumental rule 'Society's ability to respond': The ability of societies to identify and suitably react to problems in environmental and social systems arising from the energy system and to promote the energy transition has to be maintained or improved. This requires answering questions of relevance, particularly for which developments or events society should be prepared, and establishing suitable information, communication, institutional, and organizational structures.</td></tr>
<tr><td>Description</td><td>Households changing their supplier of electrical energy in order to buy 'Ökostrom' (green electricity) may support the development of renewable energy.</td></tr>
<tr><td>Unit</td><td>%</td></tr>
<tr><td>Data</td><td>

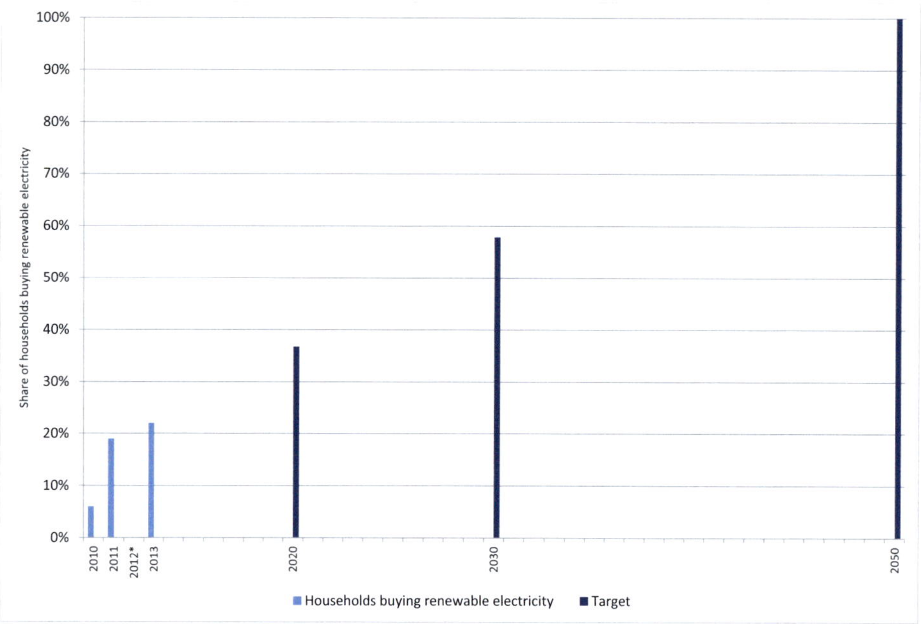

Figure 1: Percentage of households that buy renewable electricity (households with green electricity) (own diagram based on [1])

* for 2012 no data available
</td></tr>
<tr><td>Targets</td><td>2020: 37 %
2030: 58 %
2050: 100 %</td></tr>
<tr><td>Assessment</td><td>

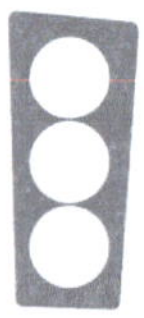

Latest available value (2013): 22 % [1]

There is no assessment possible due to a lack of data.
</td></tr>
</table>

Comments Data	Green electricity tariffs in Germany are labeled in order to provide consumers with the assurance that they get what they pay for. However, there are several labels for green electricity in Germany [2]. The so-called green electricity is not always green: What is referred to as green electricity depends on the definition in each individual case. Some green electricity certificates consider electricity from gas-fired power plants with combined heat and power as green electricity because of the relatively efficient use of energy. In other green energy certificates, only electricity generated on a renewable basis is considered as green electricity. Due to the fact that up to now only data for the general purchase of RE (whether certified or not certified) are available, Figure 1 presents the percentage of households that buy RE. Values in Figure 1 were probably gathered in a slightly different way because various surveys and literature sources were used for different years. Therefore, the numbers may not be used as a simple extrapolation (Note given in [1]). Data, based on a survey, are also given in [3] (persons in Germany, not households, who buy green electricity). According to this source, the percentage is 6 in 2010, 7 in 2011, 9 in 2012 and 11 in 2013 (share of the German population aged 14 and over).
Comments Targets	The target for 2050 should be 100 %. Data for the years 2020 and 2030 are the result of an interpolation, starting with the year 2013.
Literature	[1] Agentur für Erneuerbare Energien: Bundesländer in der Übersicht. https://www.foederal-erneuerbar.de/landesinfo/bundesland/D/kategorie/akzeptanz/auswahl/227-anteil_der_haushalte/#goto_227 , access 13 May 2015 [2] Stromtipp.de: Wieviel Öko steckt in Ökostrom-Zertifikaten? http://www.stromtip.de/rubrik2/19969/Oekostrom-Zertifikate.html, access 22 January 2015 [3] Statista: Bevölkerung in Deutschland nach Bezug von Ökostrom von 2010 bis 2016 (Personen in Millionen). http://de.statista.com/statistik/daten/studie/181628/umfrage/bezug-von-oekostrom, access 13 May 2015

41. Share of Installed Smart Meters Mandatory for Large Electricity Consumers

Justification referring to Sustainability Concept	This indicator addresses society´s ability of reflexivity. This rule demands that measures and framework conditions should be developed in a way that goes beyond addressing mere particularistic challenges or interests of actors. This requires suitable consideration of consequences of actions in thematic fields or social subsystems on others. Setting-up a information and communication tool should be one element to realize that.
Description	The indicator reveals how wide-spread smart meters for electricity are among large electricity consumers in Germany.
Unit	%
Data	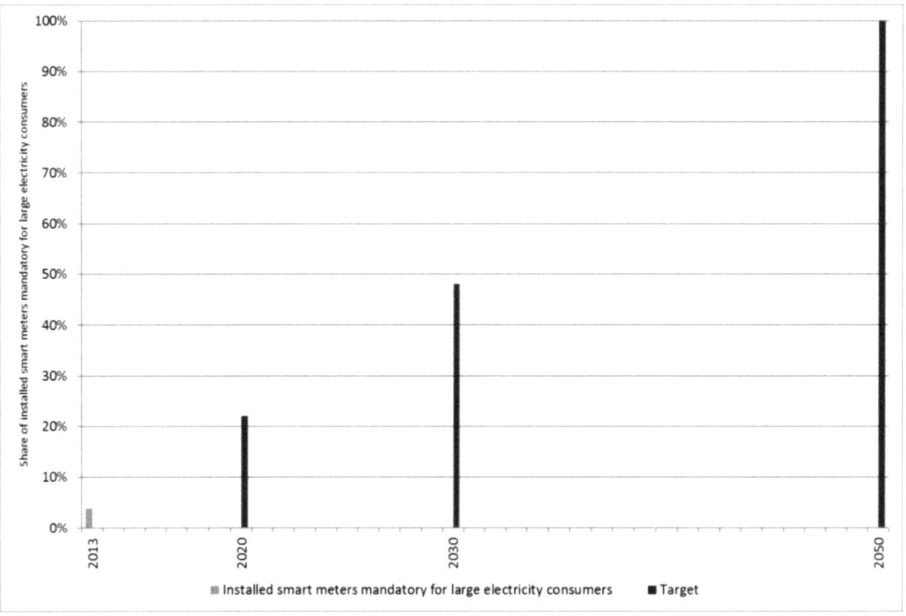 Figure 1: Share of installed smart meters mandatory for large electricity consumers (own diagram based on [1])
Targets	2020: 22 % 2030: 48 % 2050: 100 %
Assessment	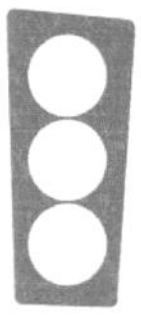 Latest available value (2013): 3.9 % [1] There is no assessment possible because only one data record is available.

Comments Data

In general, smart meters are instruments that can measure one or more of the following: heat, gas and electrical power, flow of a liquid (water or oil). Here we only reported on the number of smart meters for electricity since smart grids for electrical energy are mostly discussed as of now.

The legal status in Germany follows the advice given in the Cost Benefit Analysis by Ernst & Young for the BMWi [2]: Rollout according to the EU guideline: System fees of 29 € per year and consumer in addition to the existing annual fee of 21.60 € on average for the operation of metering points, metering and billing was found disadvantageous on the macroeconomic and the microeconomic scale.

Current legal status: The implementation of smart meters was presecribed for consumer groups which extensively strain the energy system but also have the potential to reduce their energy costs while contributing to the load transfer. These are consumers with a consumption of more than 6,000 kWh/a, measuring points in new buildings and operators of plants under the Renewable Energy Act (EEG) or the Act on Combined Heat and Power Generation (KWKG).

Implementation status in other European countries [3, 4]: Estonia, Finland, France, Ireland, Italy, Malta, the Netherlands, Norway, Portugal, Spain, Sweden, and the UK are characterized by a clear path towards a full rollout of smart metering. The mandatory rollout is either already decided or there are major pilot projects paving the way for a subsequent decision.

Share of installed smart meters according to the definition of § 21b ff. EnWG [1, p. 164].

The obligation results in § 21c: 'Metering point operators have to install measuring systems on the premises of end users whose annual consumption exceeds 6,000 kWh, ...' (own translation) [5].

The table below summarizes the number of installed intelligent measuring systems according to § 21b ff. EnWG and the measuring points that are legally obliged to install smart meters based on a survey [1, p. 164]:

	Installed	**Legally obliged**
Users with consumption over 6,000 kWh/a	171,461	4,398,207

The monitoring report 2014 gives no information about these smart meters [6]. In the monitoring report 2016, the number of measuring points required by law decreased to 4,330,915 but the number of installed meters is not given in this source [7].

Comments Targets

Large consumers like those with an annual consumption of more 6,000 KWh have a great influence on the overall demand. If there are possibilities to shift the use of energy in accordance with the necessities of the grid, they could have a positive influence on grid stability and on the need to install new power plants. Therefore, the implementation of smart meters by a large number of those consumers who are legally obliged could bring about positive effects. The precondition for realizing the potential of these positive effects is the installation of smart meters by all large consumers by 2050. The values for 2020 and 2030 were interpolated accordingly.

Literature	[1]	Bundesnetzagentur für Elektrizität, Gas, Telekommunikation, Post und Eisenbahnen; Bundeskartellamt: Monitoring report 2013. 2014. http://www.bundesnetzagentur.de/SharedDocs/Downloads/EN/BNetzA/PressSection/ReportsPublications/2013/MonitoringReport2013.pdf?__blob=publicationFile&v=10, access 3 April 2016
	[2]	Ernst & Young im Auftrag des BMWi: Kosten-Nutzen-Analyse für einen flächendeckenden Einsatz intelligenter Zähler. 2013
	[3]	Renner, S.; Albu, M.; Elburg, H.; Heinemann, C.; Lazicki, A.; Penttinen, L.; Puente, F.; Sæle, H.: European Smart Metering Landscape Report. 2011
	[4]	Hierzinger, R.; Albu, M.; Elburg, H.; Scott, A.; Lazicki, A.; Penttinen, L.; Puente, F.; Sæle, H.: European Smart Metering Landscape Report 2012 – update May 2013. 2013
	[5]	Gesetz über die Elektrizitäts- und Gasversorgung. Energiewirtschaftsgesetz vom 7. Juli 2005 (BGBl. I S. 1970, 3621), das durch Artikel 9 des Gesetzes vom 19. Februar 2016 (BGBl. I S. 254) geändert worden ist. p. 57
	[6]	Bundesnetzagentur für Elektrizität, Gas, Telekommunikation, Post und Eisenbahnen; Bundeskartellamt: Monitoring report 2014. 2014. www.bundesnetzagentur.de/DE/Sachgebiete/ElektrizitaetundGas/Unternehmen_Institutionen/DatenaustauschundMonitoring/Monitoring/Monitoringberichte/Monitoring_Berichte_node.html%20, access 3 April 2016
	[7]	Bundesnetzagentur für Elektrizität, Gas, Telekommunikation, Post und Eisenbahnen; Bundeskartellamt: Monitoring report 2016. 2017. www.bundesnetzagentur.de/SharedDocs/Downloads/DE/Sachgebiete/Energie/Unternehmen_Institutionen/DatenaustauschUndMonitoring/Monitoring/Monitoringbericht2016.pdf?__blob=publicationFile&v=2 , access 7 February 2017

42. Volume of Publicly-financed Loans for Energy-related Investments

Justification referring to Sustainability Concept	This indicator addresses the sustainability rule *'Society's capability of government'* that means that the measure has to support the capability of the government in that direction that it could help the society to develop in a more sustainable way. Based on the knowledge that in the highly complex world of today governing a society could not be in a direct way the art is to help the society to activate their own ressources and find their own solutions.
Description	This indicator reveals the overall value of loans directed at energy-related investments in private households and small and medium-sized companies.
Unit	€
Data	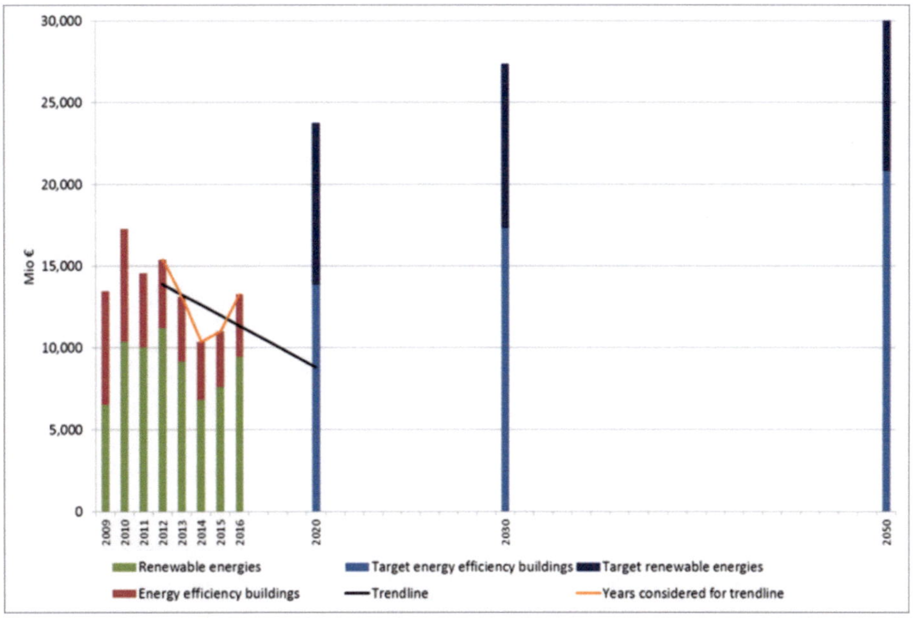 Figure 1: The total volume of Kreditanstalt für Wiederaufbau (KfW) loans for energy efficiency measures and renewable energies projects (own diagram based on [1])
Targets	2020: 23.7 billion € 2030: 27.3 billion € 2050: 31.4 billion €
Assessment	 Latest available value (2016): 13.2 billion € [1] The trend calculated based on the previous 5 years (2012–2016) shows a decrease of the loans of 30 % compared to the reference value. However, the loans should increase of 80 % in comparison to the reference value. Thus, the indicator is assessed with a red traffic light since it is very unlikely that the target of 2020 will be met.

Comments Data	The 'Förderbericht' (funding report) of KfW [1] differentiates between loans and grants (grants are not taken into consideration) and lists the Federal States in which the financial means were spent. There is no differentiation for the groups of receivers, so they may include not only private persons but also small and medium-sized companies. KfW will not provide a detailed differentiation because of bank regulations. The field of energy-related investments is divided into the funding priorities 'environment' and 'housing' in the KfW funding report. Regarding the environment, mainly activities in the field of RE and energy efficiency are funded, when it comes to housing, activities in the field of energy-efficient building and energy-efficient refurbishment are funded. The following programs were taken into consideration: Environment: 'Renewable energies – standard', 'Renewable energies – premium', 'Offshore wind energy', 'Energy efficiency program', 'Funding initiative energy transition' Housing: 'Energy-efficient refurbishment – efficient house', 'Energy-efficient refurbishment – individual measures', 'Energy-efficient refurbishment – complementary loan' and 'Remodelling housing space'.
Comments Targets	The following assumptions are made to derive target values: • For the funding priority 'environment' it is assumed that the funding amount will develop in proportion to the investments in the field of RE as listed in [2]. • For the funding priority 'housing' it is assumed that the funding will develop in proportion to the refurbishment rate of the housing stock. It is also assumed that the refurbishment rate will be constantly increasing from currently (2010) 1 % per year to 2 % per year until 2020 (see also [2]), each value based on the building stock in need of refurbishment in 2010. Our calculations are also based on the assumption that there will be a steady increase to 3 % per year until 2040 and that this value will remain unchanged until 2050. According to these presumptions, almost the complete building stock in need of renovation will be refurbished by 2050 (see also Indicator 31). • From 2022 on no loans for construction of new buildings are taken into account because by European law [3].
Literature	[1] Kreditanstalt für Wiederaufbau (KfW): KfW Förderreporte 2009 – 2016, https://www.kfw.de/KfW-Konzern/%C3%9Cber-die-KfW/Zahlen-und-Fakten/KfW-auf-einen-Blick/, access 26 April 2017 [2] DLR – Deutsches Zentrum für Luft- und Raumfahrt; IfNE – Ingenieurbüro für neue Energien; IWES – Fraunhofer Institut für Windenergie und Energiesystemtechnik: Langfristszenarien und Strategien für den Ausbau der erneuerbaren Energien in Deutschland bei Berücksichtigung der Entwicklung in Europa und global. Schlussbericht BMU, FKZ 03MAP146, March 2012. http://www.dlr.de/dlr/Portaldata/1/Resources/bilder/portal/portal_2012_1/leitstudie2011_bf.pdf, access 01 June 2017 [3] European Union 2010: DIRECTIVE 2010/31/EU OF THE EUROPEAN PARLIAMENT AND OF THE COUNCIL of 19 May 2010 on the energy performance of buildings (recast) Official Journal of the European Union 18.6.2010 L 153/13 to 35

43. Number of Energy Cooperatives Engaged in Renewable Energy Plants

Justification referring to Sustainability Concept	This indicator describes the instrumental rule *'Society's ability for self-organization'*: Potentials for the autonomous organization of societal actors who contribute to achieving the energy transition targets have to be promoted to complement top-down regulatory structures. It is crucial that these forms of civil society can be developed by using network structures and participatory decision making and can also be integrated into pluralistic and democratic planning processes.
Description	Individual citizens can get a say in the selection of their energy technology and energy production by investing in cooperative renewable energy projects.
Unit	Total number
Data	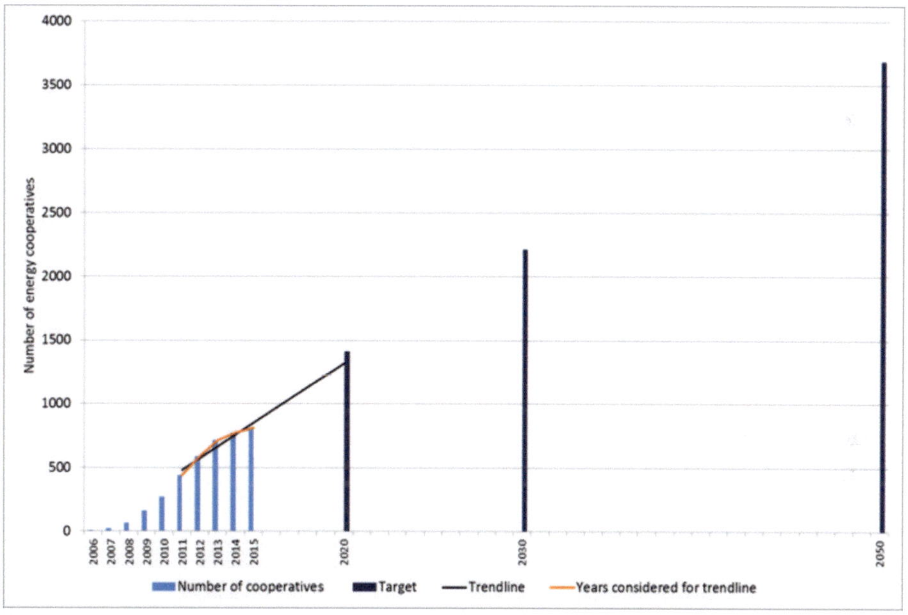 Figure 1: Total number of energy cooperatives in Germany (own diagram based on [1])
Targets	2020: 1,415 cooperatives 2030: 2,215 cooperatives 2050: 3,691 cooperatives
Assessment	 Latest available value (2015): 812 energy cooperatives [1] The trend calculated based on the previous 5 years (2011–2015) shows an increase of the total number of cooperatives of about 99 % until 2020 compared to the reference value (the average value over the years 2011 to 2015). The target value for 2020 requires an increase of 112 % compared to the reference value (666 energy cooperatives). This results in a deviation of 13 % and a yellow traffic light.

Comments Data

Only energy cooperatives under the umbrella of the DGRV (Deutscher Genossenschafts- und Raiffeisenverband e.V.) are taken into consideration. According to these studies, the number of energy cooperatives was 8 in 2006, 272 in 2010 and 812 in 2015. These figures may vary since some sources are based on the year of establishment, others on the year of registration.

The contract of these energy cooperatives includes (results are based on a survey and multiple answers were possible, see [2, 3]):

a) Electricity production at 87 % in 2012 and at 95 % in 2013

b) Heat production at 19 % in 2012 and 16 % in 2013

c) Grid operation at 4 % both in 2012 and 2013

d) Operation of district heating system at 20 % in 2012 and 16 % in 2013.

In 2011, more than 80,000 citizens were engaged in the energy cooperatives, this number increased to 136,000 in 2012 and approx. 145,000 in 2013. This means that on average each cooperative consists of 163 citizens. Civil power plants produced approx. 580 million kWh of green electricity in 2012 and approx. 830 million kWh in 2013 [2, 3].

Comments Targets

Approx. 129 energy cooperatives were established in the year 2013. After changes in the German Renewable Energy Act (EEG) this number decreased to 56 in 2014 and 40 in 2015 (see [1]). To preserve the current ability for self-organization in the field of RE, we derive the target value by assuming that the number of cooperatives will rise in proportion to the increase of the 'Share of Renewable Energy in Gross Final Consumption of Energy' (see Indicator 10 and [4]). The following table shows the respective data:

Table 1: Calculated number of energy cooperatives based on the share of RE in gross final energy consumption

	Share of RE on gross final consumption of Energy	Increase compared to 2015	Number of cooperatives
	%	Factor	
2015	13.2		812
2020	23.0	1.74	1,415
2030	36.0	2.73	2,215
2050	60.0	4.55	3,691

Literature

[1] DGRV – Deutscher Genossenschafts- und Raiffeisenverband e.V.: Energiegenossenschaften. Ergebnisse der DGRV-Umfrage (zum 31.12.2015), https://www.genossenschaften.de/sites/default/files/Auswertung%20Jahresumfrage_0.pdf, access: 23 March 2017

[2] DGRV – Deutscher Genossenschafts- und Raiffeisenverband e.V.: Energiegenossenschaften. Ergebnisse der Umfrage des DGRV und seiner Mitgliedsverbände. Frühjahr 2014, Berlin

[3] http://www.begeb.de/mediapool/139/1396625/data/Auswertung20Studie20BroschC3BCre202014.pdf, access Nov 2016

	[4] DGRV – Deutscher Genossenschafts- und Raiffeisenverband e.V.: Energiegenossenschaften. Ergebnisse der Umfrage des DGRV und seiner Mitgliedsverbände. Frühjahr 2013, Berlin https://www.dgrv.de/webde.nsf/7d5e59ec98e72442c1256e5200432395/dd9db514b5bce595c1257bb200263bbb/$FILE/Umfrageergebnisse%20Energiegenossenschaften.pdf, access Nov 2016 [5] DLR – Deutsches Zentrum für Luft- und Raumfahrt; IfNE – Ingenieurbüro für neue Energien; IWES – Fraunhofer Institut für Windenergie und Energiesystemtechnik: Langfristszenarien und Strategien für den Ausbau der erneuerbaren Energien in Deutschland bei Berücksichtigung der Entwicklung in Europa und global. Schlussbericht BMU, FKZ 03MAP146, March 2012. http://www.dlr.de/dlr/Portaldata/1/Resources/bilder/portal/portal_2012_1/leitstudie2011_bf.pdf, access 01 June 2017

44. Share of Population Living in Regions with the Objective to Shift to 100 % Renewable Energy

<table>
<tr><td>Justification referring to Sustainability Concept</td><td>This indicator describes the instrumental rule 'Society's ability for self-organization': Potentials for the autonomous organization of societal actors who contribute to achieving the energy transition targets have to be promoted to complement top-down regulatory structures. It is crucial that these forms of civil society can be developed by using network structures and participatory decision making and can also be integrated into pluralistic and democratic planning processes.</td></tr>
<tr><td>Description</td><td>Regions, cities and municipalities shifting to renewable energies are important supporters of the energy transition. Up to now, some regions have decided on the objective to shift to 100 % renewable energy.</td></tr>
<tr><td>Unit</td><td>%</td></tr>
<tr><td>Data</td><td>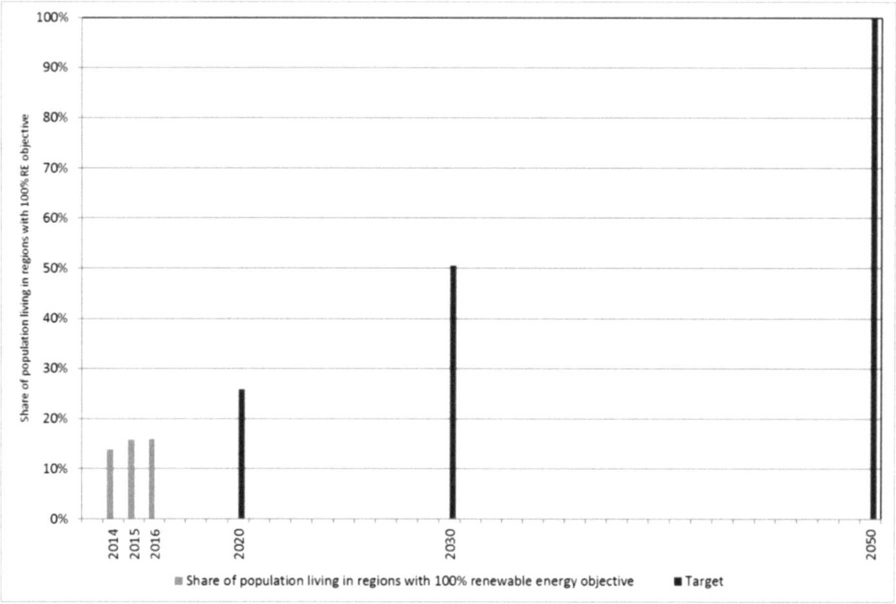

Figure 1: Share of population living in regions with the objective to shift to 100 % renewable energy (own diagram based on [1-4])</td></tr>
<tr><td>Targets</td><td>2020: 26 %
2030: 51 %
2050: 100 %</td></tr>
<tr><td>Assessment</td><td>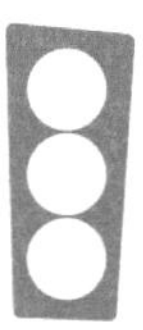
Latest available value (2016): 16 % [1],[4]
There is no assessment possible since data is only available for three years.</td></tr>
</table>

Comments Data	These data are based on the register of the 'network of regions aiming at a 100 % renewable energy supply' (100ee regions). Both rural and urban regions can register there. It is differentiated between: • 100ee regions: These regions are the forerunner of the regional energy policy. They provide space to test innovative renewable energy technologies and create new organizational forms of cooperation. They are based on a broad regional consensus on the energy transition. In addition, they have a comprehensive regional network of actors, extensive planning and conceptual groundwork, and tested tools for public relations. Their share of energy from renewable energy sources is above average. • 100ee urban: These are the pioneers of renewable energy in predominantly urban regions. In relation to other comparable urban regions, they have an above-average share of renewable energies, are planning substantial energy savings, and provide space to test innovative efficiency technologies. Furthermore, they are based on a broad consensus on energy policy, have a comprehensive network of actors, extensive planning and conceptual groundwork, and proven tools for public relations [2]. Since there is no obligation to register, it can be assumed that there are other regions, which are committed to the 100 % renewable energy supply objective but are not included here.
Comments Targets	In 2050, 60 % of the energy shall come from renewable sources [5] and according to the Paris Agreement net greenhouse gas emissions have to be cut back to zero between 2045 and 2060. To this end, according to today's state of the art, our energy supply has to be shifted to 100 % RE. Therefore the aim of all German regions until 2050 should be to shift 100 % of their energy production to 100 % renewable energies in the foreseeable future.
Literature	[1] IdE – Institut dezentrale Energietechnologien gemeinnützige GmbH: 100 % Erneuerbare-Energie-Regionen – Stand Oktober 2016. Kassel, 2016 [2] IdE – Institut dezentrale Energietechnologien gemeinnützige GmbH: 100 % Erneuerbare-Energie-Regionen – Stand Oktober 2015. Kassel, 2015 [3] IdE – Institut dezentrale Energietechnologien gemeinnützige GmbH: 100 % Erneuerbare-Energie-Regionen – Stand Oktober 2014. Kassel, 2014 [4] Statistisches Bundesamt: Pressemitteilung Nr. 033 vom 27.01.2017: Bevölkerung in Deutschland voraussichtlich auf 82,8 Millionen gestiegen. https://www.destatis.de/DE/PresseService/Presse/Pressemitteilungen/2017/01/PD17_033_12411.html, access 29 May 2017 [5] Bundesregierung: Energiekonzept für eine umweltschonende, zuverlässige und bezahlbare Energieversorgung. Berlin, 2010, p. 5

45. Share of the Four Biggest Electricity Companies on the Market for the First-time Sale of Electricity

<table>
<tr><td>Justification referring to Sustainability Concept</td><td>This indicator addresses the instrumental rule 'Balance of power between societal actors': Opinion-making, negotiation and decision-making processes related to the energy system and the energy transition must be open to societal actors to articulate their thoughts and exert influence, procedures have to be transparent and allow for broad participation. This requires, for instance, avoiding or reducing a high concentration of power, asymmetric communication and limited access to information and consultation.</td></tr>
<tr><td>Description</td><td>Market share of the four biggest electricity supply companies in Germany based on electricity volumes generated. A high share gives evidence for the accumulation of concentrated power.</td></tr>
<tr><td>Unit</td><td>Percentage of the market for the first-time sale of electricity</td></tr>
<tr><td>Data</td><td>
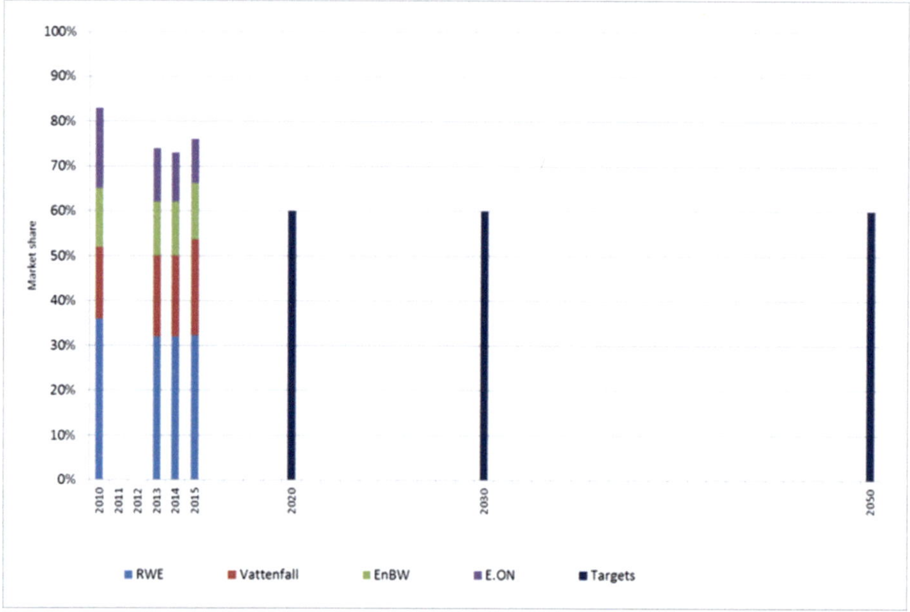

Figure 1: Shares of the four strongest suppliers on the market for the first-time sale of electricity 2010, 2013, 2014 and 2015 (own diagram based on [1-3])</td></tr>
<tr><td>Targets</td><td>2020: not more than 60 %
2030: not more than 60 %
2050: not more than 60 %</td></tr>
<tr><td>Assessment</td><td>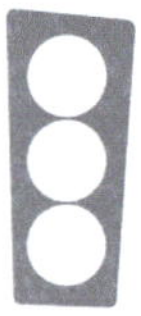
Latest available value (2015): 76 % [1]
Due to the lack of comparable data for 2011 and 2012, no assessment is possible.</td></tr>
</table>

Comments Data	The German electricity and gas markets were fully opened to competition in 1998. While there are numerous market players in both markets, neither sector is considered competitive as both are characterized by a high degree of vertical and horizontal integration and dominated by a few large companies. The situation on the electricity market has changed signficantly in the course of the energy transition and the entry of smaller companies and cooperatives producing renewable electricity. The German electricity industry is still dominated by four large electricity companies (RWE, E.ON, Vattenfall and EnBW). The companies' share generation capacities in Germany decreased from 83 % in 2010 to 74 % in 2013 and to 73 % in 2014. In 2015, there was a slight increase to around 76 %. Note that the companies' share generation capacities were calculated without EEG capacities and without capacities not connected to the general supply grid [3]. The data from [1-3] were calculated according to the dominance method which calculates the sum of the market shares of the three, four or five competitors with the largest shares in the market (CR 3, CR 4, CR 5 - 'concentration ratios') [3, p. 23]. The market definition applied to the market for the first-time sale of electricity differ to those made in the years before 2014, thus exisiting data from 2011 and 2012 are not directly comperable. However, data from 2010 were obtained from the Bundeskartellamt merger control proceeding.
Comments Targets	If concentration rates discussed in [4, p. 8] are used, which seems plausible, then market dominance is presumed if • CR 1 > 33.3 % • CR 3 > 50.0 % • CR 5 > 66.7 % i.e. the 1, 3 or 5 largest suppliers are dominating the market shares in %. In the German case with 4 significant companies, the limit shall be conservatively calculated from between the values for CR 3 and CR 5, which results in a value of about 60 %. For the corresponding target values the maximum share of 60 % for the 4 biggest electricity producing companies should be valid for the years 2020, 2030 and 2050.
Literature	[1] Bundesnetzagentur, Bundeskartellamt: Monitoringbericht 2016. November 2016, https://www.bundesnetzagentur.de/DE/Sachgebiete/ElektrizitaetundGas/Unternehmen_Institutionen/DatenaustauschundMonitoring/Monitoring/Monitoringberichte/Monitoring_Berichte_node.html, access 20 April 2017 [2] Bundesnetzagentur, Bundeskartellamt: Monitoringbericht 2015. Korrektur: 21. März 2016, https://www.bundesnetzagentur.de/DE/Sachgebiete/ElektrizitaetundGas/Unternehmen_Institutionen/DatenaustauschundMonitoring/Monitoring/Monitoringberichte/Monitoring_Berichte_node.html, access 20 April 2017 [3] Bundesnetzagentur, Bundeskartellamt: Monitoringbericht 2014. Stand 14. Nov. 2014, https://www.bundesnetzagentur.de/DE/Sachgebiete/ElektrizitaetundGas/Unternehmen_Institutionen/DatenaustauschundMonitoring/Monitoring/Monitoringberichte/Monitoring_Berichte_node.html, access 20 April 2017 [4] Lang, C.: Marktmacht und Marktmachtmessung im deutschen Großhandelsmarkt für Strom. Springer, 2008

Acknowledgements

The authors gratefully acknowledge financial support by the Helmholtz Alliance ENERGY-TRANS 'Future infrastructures for meeting energy demands – Towards sustainability and social compatibility'. The authors also thank Nora Weinberger, Benedikt Rilling, Sylke Wintzer and Miriam Miklitz.

For additional information, see https://www.energy-trans.de/